10.00

D1263543

areas of food technology and
t is becoming increasingly

DETERMINATION OF FOOD CARBOHYDRATES

DETERMINATION OF
FOOD CARBOHYDRATES

D. A. T. SOUTHGATE

Dunn Nutritional Laboratory,
University of Cambridge and Medical Research Council,
Cambridge, England

APPLIED SCIENCE PUBLISHERS LTD
LONDON

APPLIED SCIENCE PUBLISHERS LTD
RIPPLE ROAD, BARKING, ESSEX, ENGLAND

ISBN: 0 85334 693 3

WITH 19 TABLES AND 16 ILLUSTRATIONS

Printed in Great Britain by Galliard (Printers) Limited, Great Yarmouth

'those many organic substances which, tho' to
sense wholly dissimilar and incomparable in kind,
and yet all combinations of the same simples,
and even in like proportions differently disposed;'
'tis ever our old customers, carbon and hydrogen
pirouetting with oxygen in their morris antics;
the chemist booketh all of them as CHO'

The Testament of Beauty
ROBERT BRIDGES

By permission of the Oxford University Press, Oxford

Preface

In some respects the determination of carbohydrates in foods is one of the neglected areas of food analysis and this book was written in an attempt to remedy this neglect.

The idea for the book originated during a Symposium at the National College of Food Technology, Weybridge, in the Spring of 1973 when I was chairman of a study group on the subject.

Although my interest in the carbohydrates in foods is primarily nutritional, I have tried in this book to present the approach of other types of interest and in so doing to cover the interests of the practising food analyst and the food technologist.

In those areas where my direct experience was limited I have had the benefit of discussions with many analysts from different spheres of interest and I would like to acknowledge their help.

I would like also to acknowledge the helpful discussions I have had with my immediate colleagues, Mr W. J. Branch, Mr G. J. Hudson and Mrs Celia Greenberg.

My thanks are also due to Mrs Susan Moore, Miss Janice Haddock and Miss Christine Hudson for typing the various stages of the manuscript.

I would also like to thank the Association of Official Analytical Chemists and the Boehringer Corporation for their helpful approach where it was necessary to refer to descriptions of their methods. The Society of Analytical Chemists and the authors gave permission for reproduction of Fig. 6.1 from 'The Analyst'.

D. A. T. SOUTHGATE
Cambridge

Contents

CHAPTER 1

Introduction

AN APPROACH TO THE ANALYSIS OF
CARBOHYDRATES IN FOODS

The measurement of carbohydrates in foodstuffs presents special problems to the analyst. In some respects these are due to the niceties of the chemistry of carbohydrates and to the relative complexity of the mixtures of carbohydrates found in many common foods. In other respects they are a legacy from the period when carbohydrate chemistry was poorly understood and analytical methods were as a consequence empirical.

The use of empirical methods of uncertain specificity led in many specialised areas to the development of a complex and, to modern ears at least, confusing nomenclature for the carbohydrates in foods derived, in part, from the analytical methods themselves.

This combination of circumstances has led to the analytical chemistry of the carbohydrates in foods becoming separated from the main streams of carbohydrate chemistry and from biochemical and nutritional studies of the carbohydrates.

It is true that at the present time these different approaches are merging. The statutory requirements concerning the labelling of foods in the United Kingdom [111], and the impending legislation with regard to nutritional labelling in many countries, will mean that the nutritional aspects of carbohydrates will have to be more accurately reflected by the consulting analyst who, until now, has been bound by statutory analytical procedures devised in the past.

The object of this book is to examine the range of analytical procedures that have been, and are being used by analysts whose interests lie in the field of nutrition and by the practising analysts concerned with statutory and related analyses, and to review these procedures against the background of our current knowledge of carbohydrate chemistry.

The requirements of these groups of analysts are often divergent. For example, the food technologists require rapid procedures which are simple and yet highly reproducible when judged in a special way, according to the

1

final quality of his product. Such techniques will often be empirical and may be of low intrinsic accuracy.

In another area, the consulting analyst needs procedures that can be applied rapidly without the need for elaborate apparatus and with which analysts working under a variety of different conditions will obtain similar results.

In much nutritional work, on the other hand, the differences in the metabolic behaviour of structurally related carbohydrates are such that a detailed knowledge of the amounts of the different carbohydrate species present in a diet is required.

These three generalised examples illustrate the wide approach that is necessary when considering the measurement of the carbohydrates in foods, and one may feel that it is not possible to consider them together. They have, however, the unifying feature that they are all concerned with measuring the same components of the food, albeit to a different degree of precision, and for this reason they can and I feel should be discussed within the framework of one book.

THE DEVELOPMENT OF
CARBOHYDRATE ANALYSIS

The analysis of any constituent at a point of time is limited by the state of knowledge of the chemistry of that constituent. This was particularly true for the determination of carbohydrate, where the very similar properties of different sugars caused considerable difficulties. The aim of this section is to try to distinguish the various themes that have dominated the progress of analytical methods for food carbohydrates.

The early chemical exploration of the carbohydrates has been described by McCollum [75] in his book on the history of nutrition and this book is particularly valuable because it provides a useful insight into the state of knowledge of carbohydrate chemistry at different times.

In this book McCollum also describes the evolution of the chemical analysis of foods, which again serves as a useful introduction to the older literature in this area much of which is not readily accessible.

Early work on the constituents of foods was mainly carried out by the pharmacological and medicinal chemists, whose primary aim was the discovery of natural medicines. Their work led, however, to the isolation and recognition of a number of substances which would, today, be classified as carbohydrates. The crude extraction procedures used by these workers were also used analytically by a few individuals to measure the composition of vegetables and cereals.

At the same time, elementary analysis using combustion techniques was quite widely used by several distinguished workers in Europe. The later development of the classical Kjeldahl method for nitrogen by Kjeldahl in 1883 can be seen to belong to this school of food analysis. These procedures enjoyed a considerable vogue because at that time the nutritive properties of a food were thought to depend principally on its nitrogen and energy (*i.e.* carbon) content.

This view continued until Henneberg and Stohman [54] developed the Weende proximate system for the analysis of foods. This procedure, in which carbohydrate was measured by difference after deducting the moisture, protein, fat and ash from the total, and indigestible fibre was measured after extraction with hot acid and alkali, dominated the field of food analysis for nearly half a century, particularly in the large area of agricultural feed analysis. The dominance of protein and energy in nutritional circles also acted to limit interest in other constituents.

In specialised areas, however, notably in the sugar refining industry, in brewing and wine-production, considerable interest developed in the direct measurement of sugars. This interest led to a large number of methodological studies of the analysis of soluble carbohydrates by refractometry, specific gravity measurements, polarimetry and reduction methods.

These studies resulted in the accumulation of a large amount of detailed knowledge about these methods.

The methods used were, however, empirical and the need for agreement between analysts led to the establishment of an International Commission on Uniform Methods of Sugar Analysis, which first met in 1897 and has met at intervals since that time. The work of the Commission meant that in these particular areas the analyst was able to use established and well-documented, albeit empirical, procedures [128].

In general, however, the measurement of carbohydrate in foodstuffs continued to be 'by difference' with the direct measurement of fibre more or less according to the Weende method, although several efforts were made, notably by the AOAC, to refine the procedure from a technical point of view. The procedure eventually designated for measuring 'crude fibre' became embodied in feedingstuffs legislation, and led to the establishment of rigid directions for this analytical method.

A few workers attempted to measure the carbohydrates in foods directly at the beginning of this century but the results obtained using the reducing sugar methods available at the time led to highly variable results [72].

At this time there was a growth in the clinical interest in the carbohydrates in foods, primarily because this information was needed in the treatment of diabetics.

It was this interest that prompted McCance and Lawrence [72] to review the carbohydrates in foods and to attempt to measure them in a way that gave results of biological significance.

They considered that the carbohydrates in foods could be considered to fall into two broad categories. The *available carbohydrates* were those digested and absorbed by man and which were glucogenic. The *unavailable carbohydrates*, on the other hand, were not digested by the endogenous secretions of the human digestive tract.

McCance and Lawrence devised analytical procedures that they hoped would give values reflecting these nutritional differences. The procedures that they used are of largely historical interest, because Widdowson and McCance [133] showed that the procedures tended to underestimate total sugars in foods containing sucrose. However, these authors went on to develop milder and more reliable procedures for the measurement of available carbohydrates. Although their approach met with some criticism, other workers, especially Williams and his colleagues in the USA, began to measure carbohydrates directly [135] and were particularly concerned with the unsatisfactory situation with regard to the measurement of the indigestible carbohydrates [134].

These developments, however, were confined largely to the nutritional and clinical interests in food composition and appear to have had little influence on food analysis in general.

The use of two distinct methods, by analysis and 'by difference', for the determination and expression of the carbohydrates in foods led to some difficulties during World War II, primarily in connection with the assessment of the energy value of foodstuffs [79, 132]. An FAO Committee on calorie requirements [42] later recommended that the continued use of the difference method for carbohydrate was unsatisfactory and efforts should be made to evolve suitable methods for the direct measurement of carbohydrates in foods.

The advances in knowledge of carbohydrate chemistry since the end of the war and the impact of analytical biochemistry in the carbohydrate field represent the current themes in the determination of food carbohydrate and mark the point where this review of the past gives way to the present.

THEME AND GENERAL PRINCIPLES

The theme of the present work is based on the general principles involved in the complete examination of the carbohydrates in a foodstuff.

While in many cases the analyst has no need to make such an examination, the general principles involved influence the design of any shortened procedure. The use of such shortened procedures is often possible and desirable when the analyst is concerned with a familiar material or is involved in routine operations such as quality control.

Many of the procedures that appear, at the practical level, to have dispensed with many of the features of a complete examination depend on the use by the analyst of a store of received knowledge about the nature of the material under examination.

The general features of a complete scheme of analysis are given diagrammatically in Fig. 1.1 and the following chapters review in detail the various aspects of this scheme.

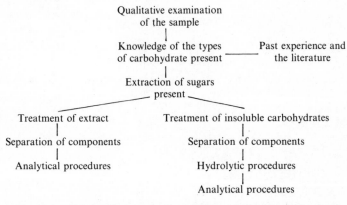

FIG. 1.1.   Schematic examination of the carbohydrates in a food.

The first stage is primarily qualitative. The wide range of carbohydrates which occurs in foods is such that the choice of the most appropriate analytical procedure can only be made intelligently, if it is based on a knowledge of the types of carbohydrate present. In many cases this knowledge can be derived from past experience as recorded in the literature and an actual qualitative examination of every sample is unnecessary.

The second chapter deals therefore with the occurrence and properties of the carbohydrates that are found in foodstuffs. The properties discussed are those of special interest and relevance to the analyst.

The third chapter deals with the extraction and analysis of the free sugars. In this chapter it is necessary to consider in some detail the separation and analysis of mixtures of sugars.

The polysaccharides in foods present a variety of different problems to the analyst and for this reason it is convenient to consider them in three separate chapters.

Chapter 4 deals with the analysis of starch and its degradation products and the structural carbohydrates of the plant cell wall are considered in Chapter 5.

The sixth chapter covers a rather heterogeneous collection of polysaccharides, the plant gums and algal polysaccharides.

Chapter 7 reviews the carbohydrates present in specific groups of foods and discusses any special analytical problems connected with those foods.

The details of selected analytical procedures are given in Chapter 8, which gives methods for specific carbohydrates or groups of carbohydrates. The Appendix contains reference tables for use with some specific methods.

CHAPTER 2

The Carbohydrates in Foods

Most chemists would have little difficulty in listing the substances which together form the carbohydrates. A precise definition is, however, more difficult. The classical definition based on the origin of the term carbohydrate, that is hydrates of carbon (L. *carbo*, Gk. *hydros*) containing hydrogen and oxygen in the proportions of 2:1, had its origins in the period when elementary analysis was the only acceptable analytical procedure for natural products and is a little naïve. Furthermore, it has the disadvantage of excluding some substances that exhibit the general reactions of carbohydrates.

The most recent definition in a legal document [111] is even more restrictive in defining them as 'any neutral polyhydroxy alcohol containing carbon, hydrogen and oxygen in which the hydrogen and oxygen occur in the same proportion as in water', but does not include any polysaccharide that is not metabolised by man.

This definition excludes many food carbohydrates that are of considerable importance in the analysis of animal feedingstuffs, in the technology of many foodstuffs and in human and animal nutrition.

Pigman and Horton [92] state that it is difficult to define the group with any degree of exactitude, and suggest two definitions. The first is more of the form of a general statement, in that 'the carbohydrates comprise several groups of homologous series characterised by the plurality of hydroxyl groups and one or more functional groups, particularly aldehyde or ketone groups, usually in the hemiacetal or acetal form'.

The second shortened statement, and in their view oversimplified, is that they (the carbohydrates) are composed of the polyhydroxy aldehydes, ketones, alcohols, acids, their simple derivatives and their polymers having polymeric linkages of the acetal type.

In this book I have preferred to use this definition because it is not restrictive and, although many derivatives of carbohydrates are not quantitatively important in foods, it seems undesirable to exclude them.

7

NUTRITIONAL CLASSIFICATION OF THE CARBOHYDRATES IN FOODS

The carbohydrates in foods have been classified in a number of different ways but one of the more important, for the worker concerned with human foods, is the system used by McCance and Lawrence [72], who considered that it was appropriate to divide the carbohydrates into two classes: *available*—those utilised and metabolised as carbohydrate, which includes starch and the soluble sugars (sucrose, glucose and fructose) and *unavailable*—those broken down extensively, but as a rule incompletely, by symbiotic bacteria, yielding fatty acids and thus not supplying the host with carbohydrate. This includes the hemicelluloses, fibre and the other complex polysaccharides. The lignin component of the fibre was mentioned briefly and it was regarded as wholly indigestible, but, presumably because it was not a carbohydrate, it was not considered to any great extent.

This definition and classification still has some value and, although knowledge of polysaccharide chemistry has advanced considerably since 1929, some of their observations remain pertinent.

This classification of the carbohydrates in foods can be illustrated schematically as shown in Fig. 2.1.

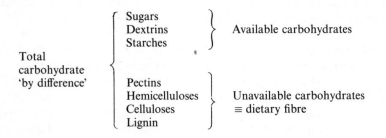

FIG. 2.1.   Classification of carbohydrates in foods [105].

The unavailable carbohydrates as defined by McCance and Lawrence [72] and as subsequently measured by McCance *et al.* [74] are identical with the term Dietary Fibre introduced by Trowell [118, 107].

This classification is useful nutritionally but for analytical purposes a more detailed and chemically defined classification is essential.

ANALYTICAL CLASSIFICATION OF THE CARBOHYDRATES IN FOODS

The carbohydrates occurring in foods are set out in Table 2.1, which gives the systematic [57] and trivial nomenclature for the various carbohydrate species.

In this table the carbohydrates are grouped under two broad headings, *Free sugars* and *Polysaccharides*. The table also gives some indication of the relative importance of the different species in a typical mixed diet eaten in Great Britain. For most practical purposes, the systematic terms are not necessary and throughout this book only the trivial names will be used, unless systematic nomenclature is required. Foods usually contain a mixture of two or more different carbohydrates and it is often possible for the analyst to adapt the selected analytical method for a particular foodstuff. This approach will be described in the chapters dealing with specific foodstuffs but before this is done it is necessary to consider all the carbohydrates that would be present in a typical diet.

Free Sugars
These are often present in solution in a plant food or for example a beverage. The most important monosaccharides are the hexoses, glucose and fructose; the pentoses, arabinose and xylose are occasionally seen but in most diets they are relatively rare. Galactose is also rarely seen, except where fermented milk products such as yoghurt form part of the diet, and even in these products the concentrations of free galactose are low.

Sucrose is the most important disaccharide and both lactose and maltose are relatively minor components in a mixed diet. In milk products, lactose is often the only carbohydrate present and many manufactured foods contain glucose syrups, which provide appreciable amounts of maltose and its higher homologues.

Trisaccharides, such as raffinose, are relatively rare components of a mixed diet but in specific foods they are present in significant quantities and it is in these specific foods that they concern the analyst.

This is also true for a number of carbohydrate species not included in Table 2.1—for example the Avocado pear contains the heptose sugar mannoheptulose and small amounts of other higher monosaccharides. Traces of unusual sugars have frequently been reported in other plants. However, most of these are not normal items in the diet.

The analyst embarking on the analysis of an unfamiliar plant food, or indeed of any unfamiliar foodstuff, must first establish the particular

TABLE 2.1

CARBOHYDRATES IN FOODS (a) FREE SUGARS

General grouping	Carbohydrate Species Represented			Relative importance in diet[b]
	Class[a]	Individual species[a]	Systematic nomenclature	
Free sugars	Pentoses	Arabinose	L-Arabinose	Rare
		Xylose	D-Xylose	Rare
	Hexoses	Glucose	D-Glucose	Major
		Fructose	D-Fructose	Major
		Galactose	D-Galactose	Rare
		Sucrose	β-D-Fructofuranosyl α-D-glucopyranoside	Major
	Disaccharides	Lactose	O-β-D-Galactopyranosyl-(1→4)-α-D-glucopyranose	Minor[c]
		Maltose	O-α-D-Glucopyranosyl(1→4)-α-D glucopyranose	Minor[c]
		Raffinose	O-α-D-Galactopyranosyl-(1→6)-α-D-glucopyranosyl β-D-fructofuranoside	Rare
	Oligosaccharides	Maltotriose and higher homologues	O-α-D-Glucopyranosyl(1→4)α-D gluco-pyranosyl(1→4)-α-D-glucopyranose	Minor[c]

[a] This list is not exhaustive.
[b] This column refers to the relative importance in a typical British diet and is very approximate.
[c] The importance of these carbohydrates in a diet is very dependent on the foods making up the diet.

TABLE 2.1—*continued*
CARBOHYDRATES IN FOODS (b) POLYSACCHARIDES

General grouping	Major classes	Types present	Chemical structural features	Relative importance in diet
Reserve polysaccharides	Starches	(Dextrins)	Partially hydrolysed starches of variable composition	Minor
		Amylose	Linear 1→4 α-glucan	Major
		Amylopectin	Branched 1→4, 1→6-α-glucan	Major
		Modified, cross-linked starches	Various types of ether or ester cross-linkings	Minor
	Other reserve polysaccharides	Glycogen	Branched 1→4, 1→6 α-glucan	Minor
		Fructans, *e.g.* Inulin		Minor
		Mannans	Linear mannans	Minor
		Glucomannans	Linear	Minor
Structural polysaccharides	Non-cellulosic polysaccharides	Pectic substances	Galacturonans, Galacturonorhamnans, Arabinogalactans and Arabinans	Minor
		Hemicelluloses	Xylans, linear and branched molecules with glucuronic acid and or arabinose sidechains Galactomannans Arabino-galactans	Minor
		Gums Mucilages	Wide range of complex heteropolysaccharides	Minor
	Cellulose	Varying degree of polymerisation	Linear 1→4 β-glucans often containing traces of other sugars	Minor
Algal polysaccharides	Sulphated	Carragenan Agar	Complex polymers containing galactose, dehydrogalactose and ester sulphate groups	Minor
	Unsulphated	Alginates	Complex polymers of guluronic and mannuronic acids	Minor

carbohydrate species present in the food by a suitable qualitative examination.

Polysaccharides

These fall into three major groupings, Reserve polysaccharides, Structural polysaccharides and Algal polysaccharides. The division between the first two groupings is admittedly arbitrary and the inclusion of the gums and mucilages in the structural category is based on the grounds that these polysaccharides contain related chemical structural features.

Reserve Polysaccharides

The most important type of polysaccharide present in this category as far as the food analyst is concerned is starch, although possibly 'the starches' would be a more correct designation. The chemically modified starches are also in this group, as are the partial hydrolysis products of starch, the dextrins. Structurally two types of molecule are present, the linear 1-4α-linked amylose and the 1,4α- and 1,6α-branched polymer amylopectin. The proportions of these two components will vary according to the foods in the diet but together they constitute the major part (80–90 %) of the total polysaccharides in most diets. The other reserve polysaccharides are minor components of the diet but in specific foods or products are of concern to the analyst.

Structural Polysaccharides

This includes a wide range of different polysaccharides; of these cellulose is perhaps the best defined, as a linear 1,4-β-glucan with a high degree of polymerisation. The non-cellulosic polysaccharides form a very heterogeneous grouping and include the, in the main, water-soluble, pectic substances and the alkali-soluble hemicelluloses.

Many common features of chemical structure are shared by these two classes of polysaccharide and I feel that, as far as the analyst is concerned, it is most useful to think of them as one group rather than to separate them arbitrarily. Gums and mucilages also fall into this group for related reasons.

Algal Polysaccharides [89]

Algal polysaccharides, notably alginates, and the sulphated polysaccharides of the carrageenan and agar types are frequently used as food additives and therefore must also be considered as food carbohydrates.

Individually the non-cellulosic and algal polysaccharides are very minor components of a mixed diet, although in some specific foods they are present in significant concentrations. Furthermore, there is increasing evidence that the amount of unavailable carbohydrates or Dietary Fibre present in the diet may be of considerable importance in clinical nutrition [107].

PHYSICAL AND CHEMICAL PROPERTIES OF CARBOHYDRATES RELEVANT TO THEIR ANALYSIS

Free Sugars
The analysis of food carbohydrate has traditionally depended on the use of both physical and chemical properties. The physical properties provided on one hand a variety of rapid procedures for estimating the concentrations of sugars in solution and, on the other hand, specific procedures for sugars in solution using polarimetry.

In part this reliance on physical properties was necessary because the closely related chemical structures of the monosaccharides limited the specificity of the earlier chemical analytical procedures.

The food analyst now has the opportunity of using the specific biochemical properties of the carbohydrates as another approach to their determination.

Physical Properties
Some physical properties of the free sugars found in foods or as components of polysaccharides, are given in Table 2.2. The table gives the molecular weights, melting point, specific rotation and some limited solubility data. The molecular weights of the hydrated carbohydrate are given if this is the most commonly available form.

The melting point data are rather imprecise, because carbohydrates tend to decompose before or around their melting point. The specific rotation data are often given in compilations in the literature in a variety of ways and it has been difficult to avoid this in the table.

The solubility data for many of the less common sugars are rather sparse and are frequently quoted in an unusual fashion.

The general features of the free sugars found in foods can be summarised as follows. In the main they are very soluble in water, some extremely soluble indeed, they are less soluble in alcohols (ethanol or methanol) but all are soluble to some extent in hot aqueous alcohols. Their solubility in non-polar organic solvents such as diethyl ether and benzene is virtually negligible, although some have an appreciable solubility in pyridine.

TABLE 2.2

SOME PHYSICAL CHARACTERISTICS OF THE COMPONENT SUGARS OF FOOD
CARBOHYDRATES [33, 81, 88]

	Molecular weight	*Melting point (°C)*	*Specific rotation [α]20–25 D*	*Solubility*		
				Water	*Alcohol*	*Ether*
Pentoses						
L-Arabinose	150·13	160	+ 105 m	v	p	i
D-Xylose	150·13	145	+ 19 m	v	s	p
Hexoses						
D-Glucose	180·16	146	+ 52·7 m	s	s	i
monohydrate	198·17	83–6		v	p	i
D-Galactose	180·16	165	+ 80·2	v^h	p	i
D-Mannose	180·16	132–3	+ 14·6 m	v	p	i
D-Fructose	180·16	102–4	− 92·4 m	v	p	i
6-Deoxyaldohexoses						
L-Fucose	164·16	145	− 76·3	v	p	i
L-Rhamnose	164·16	94 (hydrate)	+ 8·2	v	p	i
Uronic acids						
D-Galacturonic	194·14	159–60	+ 50·9 m	s	s^h	i
D-Glucuronic	194·14	165	+ 36·3 m	s	s	i
D-Mannuronic	194·14	120–30	− 6 m	s	p	i

m = solutions show mutarotation; v = very soluble; s = soluble; p = partially
soluble; i = insoluble; h = hot solvent.

Solubility as an Analytical Property

Density. The fact that it is possible to prepare very strong aqueous
solutions of many of the free sugars makes the use of specific gravity or
density measurements of sugar solutions an extremely valuable rapid
method of analysis for estimating the strengths of these solutions. A
number of systems have been evolved for relating specific gravity to
concentration of sugar. These methods have found their greatest use in the
measurement of the strengths of sucrose syrups.

The relationship between specific gravity measured at 20/20 °C and the
concentration of sucrose solutions by weight and the two common scales
used in this type of work is shown in Fig. 2.2. The method of measurement
can be by hydrometer, either calibrated directly in °Brix or the Baumé
modulus, or by weighing using a pycnometer. A detailed description of the
procedure is given in Chapter 8.

Refractive index. The refractive index of a clear solution of sugars (and many other solutes with similar solubilities) can also be used to determine the strength of solutions. The relationship is illustrated in Fig. 2.3 and details of a procedure are given in Chapter 8.

Optical rotation. The optical activity of the sugars provides a method for measurement of the concentrations of sugars in solution. The method

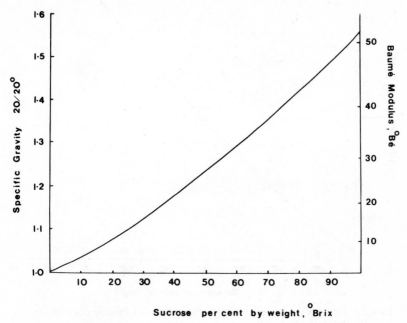

FIG. 2.2. Relationship between specific gravity and sucrose concentration.

depends for its accuracy on there being no other optically active species present. Methods based on polarimetric measurements have been widely used in the measurement of sucrose in the presence of other sugars, particularly the products of inversion.

The method also has some considerable value in determining the precise strength of standard sugar solutions where it is not advisable or possible to dry the sugar completely.

Its principal use as an analytical method lies in the measurement of sucrose, and this is described in Chapter 8 (Polarimetric methods).

Polarimetric methods have been used widely for the analysis of starch in cereals and these are also discussed later (*see* Chapter 8).

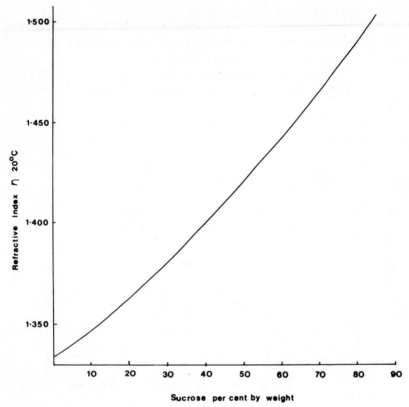

FIG. 2.3.   Relationship between refractive index and sucrose concentration.

Chemical Properties
The monosaccharides found in the carbohydrates in foodstuffs are
principally members of the D series of aldoses and ketoses and the structural
relationships within these two groups are shown in Figs. 2.4 and 2.5. In these
figures the linear form (Fischer projection) of the molecule is shown to
emphasise the structural relationships; however, in subsequent figures the
ring forms are used. In these by convention (Haworth) groups that fall to the
right in the Fischer projection are shown below the plane of the ring. The
anomeric hydroxyls are shown below the ring in α-forms of D series and
above for the α-forms of L series.

In the D-aldose series three hexoses (glucose, mannose and galactose) are
important constituents of the polysaccharides in foods or occur free in

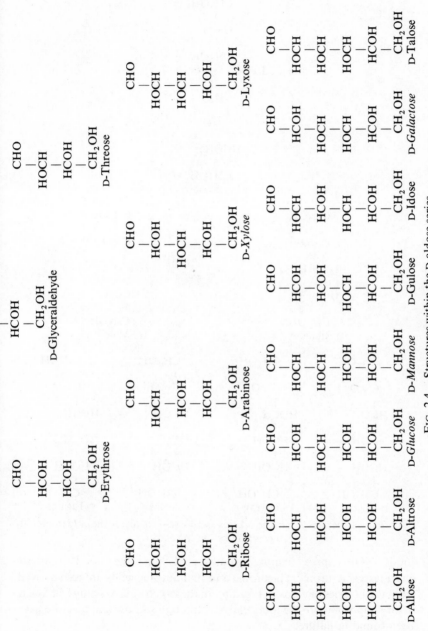

Fig. 2.4.   Structures within the D-aldose series.

(Monosaccharides set in italic are present in many foods, either free or combined in polysaccharides.)

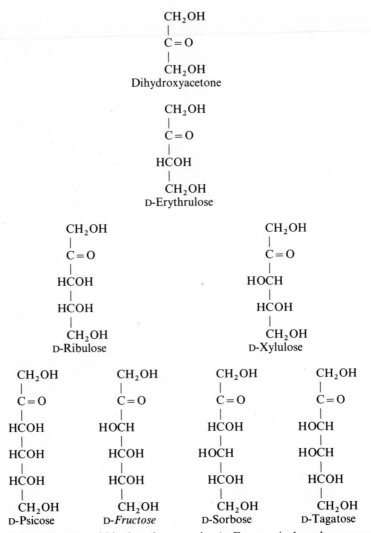

FIG. 2.5.   Structures within the D-ketose series. (D-Fructose is the only commonly occurring ketose in foods.)

foods. The three major uronic acids in the food polysaccharides: galacturonic, glucuronic and mannuronic acids are also derived from the D series of aldoses (Fig. 2.6). In the D-ketose series only fructose is of any importance to the food analyst, although sorbose and tagatose have been found in nature.

The sugars derived from the L series are fewer in number. L-Arabinose is an important component of many polysaccharides, as are the two deoxy sugars L-rhamnose (6-deoxy-L-mannose) and L-fucose (6-deoxy-L-galactose) (Fig. 2.7). L-Galactose is also an important component of some algal polysaccharides, as is L-guluronic acid.

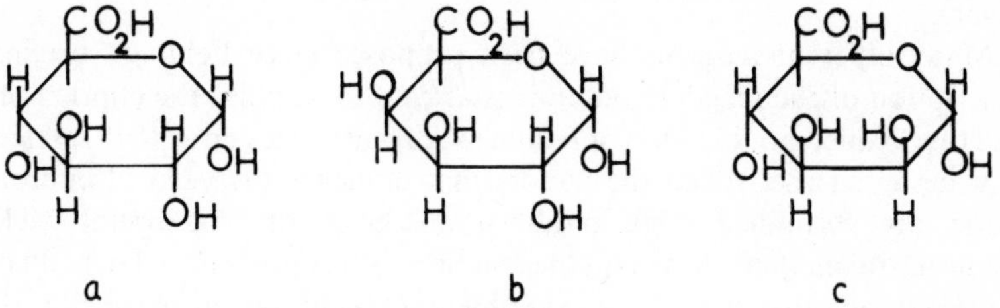

	a		b		c

FIG. 2.6.   Structures of uronic acids. (a) α-D-glucuronic acid; (b) α-D-galacturonic; (c) α-D-mannuronic.

The chemical properties of the monosaccharide sugars that are of principal importance to the analyst can be considered under a number of headings.

Reducing properties of the aldehyde or keto group. This is probably the most important chemical property that has been used in sugar analysis. The free monosaccharides will reduce alkaline solutions of metallic salts to the oxide or the free metal. The analyst has made most use of this property with regard to copper salts, and many procedures are based on the measurement of copper oxide formed in this reaction.

The reaction between the sugar and cupric sulphate does not appear to be stoichiometric and the yield of cuprous oxide is very dependent on the reaction conditions. The major factors that affect the reaction are the rate of heating, the alkalinity and the strength of the reagent. The widespread use of copper reduction methods in sugar analysis is a tribute to the patience of

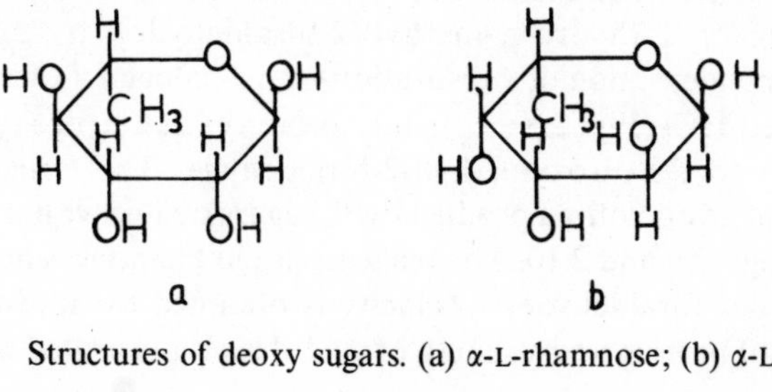

	a		b	

FIG. 2.7.   Structures of deoxy sugars. (a) α-L-rhamnose; (b) α-L-fucose.

the many chemists who examined the empirical nature of the reaction and were thereby able to devise reproducible and accurate methods. Pigman and Goepp [91] when describing these methods express an element of surprise in that such an intrinsically unpromising reaction could become the basis of reliable quantitative methods. These authors also give a detailed account of the early history of the development of copper reduction methods for sugars.

Many different reagents have been proposed since Fehling's original description of the reagent containing tartrate to stabilise the cupric ions, and the Soxhlet modification of Fehling's reagent suggested in 1878 remains in widespread use. When the factors that influence the yield of cuprous oxide are examined, this modification gives the maximum yield. Quantitative methods based on the reaction with cupric salts all depend on careful standardisation of the experimental conditions, in particular, the rate of heating, The empirical standardisation of reduction against sugar content forms the basis of methods using this principle and these are described in detail in Chapter 8 (Reducing sugar methods).

The reduction of ferricyanide has also formed the basis of quantitative methods, and again, these depend for their reliability on a careful control of the experimental conditions.

Condensation reactions. When heated in strong acid the monosaccharides yield substances that undergo condensation reactions with a range of phenolic and other substances. The reaction is believed to be a cyclisation involving loss of water and the formation of furfural derivatives. The yield of chromogen is very dependent on the precise experimental conditions; careful control of acid strength and the time and rate of heating is necessary if the reaction is to form the basis of a quantitative method. A variety of products can be isolated from the reaction mixture of acids with many individual sugars.

In the case of pentoses and methyl pentoses, the products are 2-furfurals (Fig. 2.8), which are relatively stable under acid conditions and can be distilled from the reaction mixture. The initial product of the reaction from hexose sugars [5(hydroxy-methyl)-2-furaldehyde] is unstable and undergoes condensation or substitution with a halogen if the reaction is in concentrated HCl (Fig. 2.8). Fructose produces 2,(2-hydroxyacetyl)furan in addition to 5(hydroxymethyl)-2-furaldehyde. The structures of the reagents that give coloured products with sugars treated with strong acid are shown in Figs. 2.9 and 2.10. The reagents are in themselves not specific for any particular furaldehyde; specificity is obtained by adjustment of the conditions of heating and strength of acid. This is illustrated by comparing

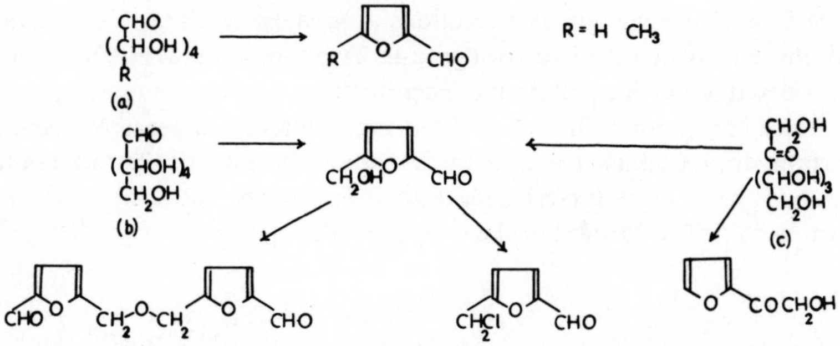

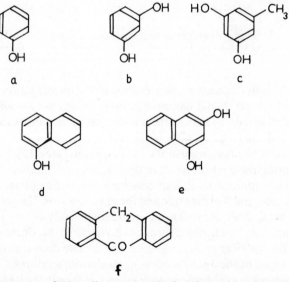

Fig. 2.8. Formation of furans. (a) Pentoses and methyl pentoses; (b) aldohexoses; (c) ketohexoses.

the conditions used with a range of reagents by Dische [35] and summarised in Table 2.3.

The actual nature of the coloured complexes formed has not been established for the majority of these reactions, and the conditions for using them in quantitative work are therefore established empirically.

Fig. 2.9. Structures of phenolic compounds forming coloured products with sugars. (a) phenol; (b) resorcinol; (c) orcinol; (d) α-naphthol; (e) naphthoresorcinol; (f) anthrone.

Substitution reactions. The reactions of greatest concern to the analyst are those involving the hydroxyl groups. Two types of reaction need to be considered, ether formation and esterification.

Ether formation. The more important of these in terms of general carbohydrate chemistry are the methyl ethers, because of their role in the establishment of polysaccharide structures. Jones and Hay [59] have written a concise account of the general methods of preparing ethers. The

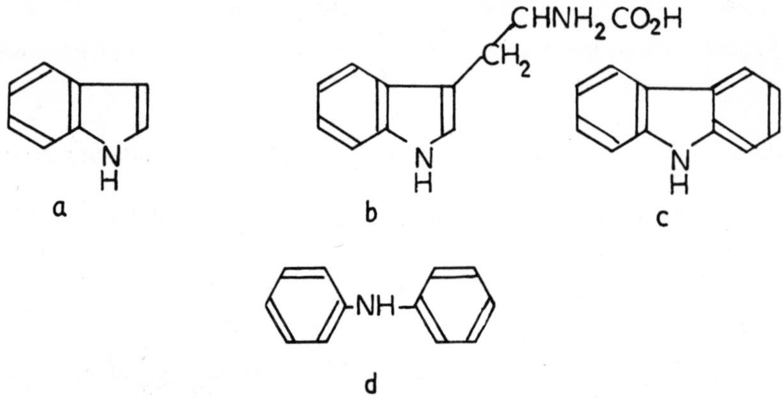

FIG. 2.10.　Structures of nitrogenous compounds forming coloured products with sugars. (a) indole; (b) tryptophan; (c) carbazole; (d) diphenylamine.

analyst's primary interest in etherification lies in the preparation of volatile derivatives for gas–liquid chromatography. Satisfactory resolution of the fully methylated derivatives of the monosaccharides and disaccharides can be achieved, and the trimethylsisyl ethers have also provided suitable derivatives for the analysis of monosaccharide mixtures [15]. Most methods for preparing these ethers result in the formation of the various anomeric forms and, in some cases, all four possible methylated anomers are formed, *e.g.* from the α and β pyranoside and furanoside forms. This can reduce the value of these ether derivatives in quantitative analysis.

Esterification. The hydroxyl groups also undergo esterification reactions with acid anhydrides or acylchlorides. The acetate derivatives have proved of some value in the analysis of monosaccharide mixtures.

The main analytical application of esters has been in the separation of the alditol acetates, where the monosaccharide sugars are reduced to the corresponding alditols, which are then acetylated. The alditols do not exist

TABLE 2.3
COMPARISON OF CONDITIONS FOR COLOURED-COMPLEX FORMATION

Reagent	Acid	Strength in reaction medium (% v/v)	Time (min)	Temp (°C)	Sugars reacting	Colour of complex	Absorption maximum (nm)
Indole	H_2SO_4	68	10	100	All	Brown, varies with class of sugar	470
α-Naphthol	H_2SO_4	80	3	100	All	Purple	550, pentoses 560, methylpentoses 570, hexoses
Tryptophan	H_2SO_4	67	20	100	All	Violet-brown	500
Cysteine	H_2SO_4	80		Room temp.	All	Yellow	390, pentoses 400, methylpentoses 463, hexoses
Anthrone	H_2SO_4	67	16	100	Hexose > > pentoses	Blue	625
Diphenylamine	Acetic/ HCl	67	10	100	Ketoses > > > aldoses	Blue	635
Resorcinol	HCl	10	10	80	Ketoses > > > others	Purple	515
Carbazole	H_2SO_4	80	10	100	All	Purple-brown	490, heptoses 525, pentoses 535, hexoses
	H_2SO_4/ borate	87	20	100	Uronic acids	Purple	515
Orcinol	HCl	18	45	100	Pentoses > > hexoses	Green	670

as cyclic structures and therefore the alditol acetates represent, in general, simpler mixtures to resolve.

Complex formation. Sugars in boric acid solution show enhanced optical rotations; this and electrophoretic data show that sugars form borate complexes, although few crystalline borate esters have been isolated [70]. Despite this, complex formation with borate is widely used as the basis for the ion-exchange separation of sugar mixtures and polysaccharides by electrophoretic methods.

Polysaccharides

A large number of different types of polysaccharide occur in foods but their physical properties have not been used to any great extent in food analysis.

Solubility in Aqueous Reagents

Many food polysaccharides are soluble in water, particularly on heating; dextrins dissolve quite readily and the amylose fraction of starch forms colloidal aqueous solutions. Amylopectin and glycogen are only partially dispersible in hot water. Starch granules absorb water on heating and swell in the process of gelatinisation. Many glucofructans are readily water-soluble on heating and will undergo some degree of autohydrolysis in aqueous solution.

Pectic substances are characteristically soluble in hot water unless they are complexed with calcium. The solubility of the other non-cellulosic polysaccharides ranges from being freely soluble in water—many gums and mucilages, the arabinoxylans and β-glucans of cereals—to those virtually insoluble in water, such as cellulose, some mannans and xylans. Solubility of the non-cellulosic polysaccharides is often determined by the structural type, highly branched molecules tending to be more soluble than the linear parent backbone. This is particularly marked in the mannans and to some extent in the xylans, where the linear unsubstituted molecules are only soluble in strong alkali.

The hemicelluloses are, by definition, insoluble in water but dissolve in alkali; here again solubility depends on the side chains and the less substituted molecules tend to require stronger alkali to bring them into solution.

Cellulose in its native state is virtually insoluble in water and the so-called α-cellulose is insoluble in 17·5 % (w/v) NaOH. Substituted celluloses, *e.g.* methyl-cellulose, dissolve in water and relative solubility depends on the extent of the substitution. Microcrystalline celluloses that will form stable colloidal suspensions in water can be prepared by physical means such as prolonged ball-milling but this is believed to result in some degree of depolymerisation.

Cellulose will also dissolve in cuprammonium salt solutions, from which it can be reprecipitated by acidification. The process of solution involves a considerable degree of depolymerisation and the reprecipitated cellulose has a different structure to the native cellulose in the plant cell wall.

In general, the polysaccharides as a class are insoluble in aqueous alcohols at strengths above 75–80 % (v/v). Many are precipitated at lower strengths but some arabinans and the glucofructans, [80] in particular, are soluble at alcohol concentrations around 70 % (v/v).

Fractional precipitation from aqueous solutions provides a method for fractionating some polysaccharide mixtures and extraction with aqueous alcohol forms the basis of many schemes for separating free sugars from polysaccharides.

Chemical Properties
In general, these have not been exploited by the analyst and the two properties that have been used are complex formation and hydrolysis.

Complex formation. Many polysaccharides form insoluble complexes with inorganic reagents and these have been used analytically in the isolation of specific polysaccharides or polysaccharide types.

Of the complex-forming reagents, iodine has attracted the most attention. Starch forms an insoluble complex with iodine in NaCl/KI solution [50]. Many linear polysaccharides of the non-cellulosic class also form insoluble iodine complexes in $CaCl_2$ solution, and this has been used as a procedure for separating linear from non-linear polymers in extracts from plant tissues [46].

Some types of xylan also form precipitates with Fehling's solution but in general this property appears to have been used entirely empirically in the fractionation of cell wall polysaccharides and its analytical use has been limited.

Sulphated algal polysaccharides of the carrageenan type will form complexes with quaternary ammonium compounds and, in some cases, these complexes are useful in fractionating these polysaccharides.

Hydrolysis. Hydrolysis, usually with mineral acids, has been used very widely as a means of analysis and the polysaccharides have been measured as their component monosaccharides.

These components with the exception of fructose and possibly arabinose are reasonably stable in dilute acid solution. Slow destruction of acid

(a) *Glucose*

glucose + ATP $\xrightarrow[hexokinase]{}$ glucose-6-phosphate + ADP

glucose-6-phosphate + NADP $\xrightarrow[glucose-6-phosphodehydrogenase]{}$ 6-phosphogluconate + NADPH + H$^+$

(b) *Fructose*

fructose + ATP $\xrightarrow[hexokinase]{}$ fructose-6-phosphate + ADP

fructose-6-phosphate $\xrightarrow[phosphoglucose\ isomerase]{}$ glucose-6-phosphate

(c) *Lactose*

lactose + H$_2$O $\xrightarrow[\beta\text{-}galactosidase]{}$ glucose + β-galactose

galactose + NAD $\xrightarrow[galactose\text{-}dehydrogenase]{}$ galactonolactone + NADH

(d) *Maltose*

maltose + H$_2$O $\xrightarrow[\alpha\text{-}glucosidase]{}$ 2- glucose

(e) *Sucrose*

sucrose + H$_2$O $\xrightarrow[\beta\text{-}fructosidase]{}$ glucose + fructose

FIG. 2.11.   Biochemical reactions used in the analysis of sugars.

solutions of most monosaccharides will occur on prolonged heating, particularly if protein or amino acids are present. In general, however, most of the polysaccharides can be hydrolysed with good yields of their component monosaccharides by dilute acid hydrolysis. Cellulose is rather resistant to dilute acid hydrolysis unless it has been partially degraded by bringing it into solution in strong acid reagents such as 72 % (w/w) H_2SO_4.

The arabinofuranosyl linkage is particularly susceptible to acid hydrolysis and release of these groups will occur in 0·01N acid at room temperature. Many glucofructans and fructans such as inulin also undergo some autohydrolysis at pH values not much on the acid side of neutral.

The presence of uronic acid residues in a polysaccharide confers great stability on the adjacent glycosidic bond and most uronic acid residues are released in combination with another residue as aldobiuronic acids. These disaccharides are extremely resistant to acid hydrolysis and can only be hydrolysed in conditions which cause considerable degradation of the component sugars and decarboxylation of the uronic acids themselves.

Biochemical Properties
The biochemical reactions of the monosaccharides and the hydrolytic enzymes acting on oligo- and polysaccharides provide the food analyst with a range of highly specific reactions, which have recently become of considerable importance [13, 20]. Advances in the techniques for the separation and purification of enzymes and a wide range of biochemical intermediates, coupled with their production on a commercial scale, have meant that these products can be used in routine analytical work and can provide the analyst with a range of specific and elegant procedures.

The basis of these biochemical reactions for analysis *per se*, involves the use of reactions leading to the production of reduced nucleotides, which can be measured by following changes in the specific absorbance at 340 or 360 nm [20]. The major reactions that have been used as the basis for analytical methods are set out in Fig. 2.11. Details of the procedures are described in Chapter 8 (Biochemical procedures).

The hydrolytic enzymes also have an important role in the evolution of specific methods for analysis of oligosaccharides and polysaccharides and they have a great potential in the analysis of food carbohydrates.

CHAPTER 3

The Measurement of Sugars

This chapter deals with the measurement of the free sugars in a food. As such it is usually concerned with the measurement of a mixture of monosaccharides, disaccharides and, to a lesser extent, tri- and higher oligosaccharides, the latter becoming more important where foods containing glucose syrups are concerned.

The analyst may be concerned with a food containing only one particular species of sugars, for example, in the analysis of milk and milk products or of a sucrose syrup, but will more usually be concerned with the analysis of a mixture.

For many purposes a value for total free sugars, expressed possibly in terms of a monosaccharide, may be sufficient, in other cases a detailed analysis of the various carbohydrate species may be required.

The analyst therefore needs a selection of procedures at his disposal if he is to be able to deal with all these possible requirements.

If only one species of sugar is present, then the choice of procedures is to a great extent much simplified, and a relatively non-specific procedure such as reducing method can be used, provided that calibration tables are available for the particular sugar in question. If a mixture is present, then the approach will depend on the need for values for total free sugars or whether separation or some other method is appropriate for the analysis of the mixture.

Thus the analyst needs a series of procedures of graded complexity depending on the nature of the sample and the analytical values required. The options are set out in Fig. 3.1.

PREPARATION OF THE SAMPLE FOR ANALYSIS

The first stage in the analysis of the free sugars depends primarily on whether the sugars are in solution or whether they need to be extracted from the foodstuff.

Samples where the Sugars are in Solution

The main requirement in the treatment of these samples is the removal of substances likely to interfere with the subsequent measurements. These may be considered under the following headings

Dissolved Gases

Carbonated or fermented products must have the dissolved gas removed; this can be done by repeatedly pouring the sample from one container to another or, more conveniently, by the application of reduced pressure to the sample in a suitable container.

Pigments and Other Colouring Matters

These will interfere with colorimetric procedures and frequently with other methods, particularly those depending on reducing reactions. A wide range of procedures has been used, ranging from decolorisation with charcoal to the range of procedures based on treatment with lead salts.

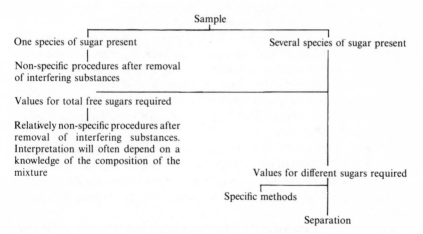

FIG. 3.1. Options in the analysis of free sugars.

The precautions relating to the use of lead salts, decolorisation or defaecation procedures are outlined in the AOAC Official methods of analysis [4] and may be summarised as follows. Basic lead acetate must not be used where polarimetric measurements are intended and, in general, neutral lead acetate is the preferred form. It is used as a saturated aqueous solution and excess lead is removed with sodium or potassium oxalate.

Deproteinisation

Proteins will also interfere in both reducing and colorimetric sugar methods, and it is usually desirable to remove the bulk of the protein before attempting analysis. Care is necessary in the choice of deproteinising agents; strongly acid reagents such as trichloroacetic acid can obviously not be employed in the treatment of solutions containing disaccharides or fructose. Two approaches seem to be acceptable.

First, the addition of an organic solvent such as ethanol or acetone to a concentration of around 70 % (v/v), at which most proteins coagulate and can be removed by centrifugation or filtration. The second approach involves the formation of a bulky, usually, inorganic precipitate to act as a collecting precipitate on which the protein forms. Ferric hydroxide has formed the basis for the deproteinisation of milks prior to sugar analysis. The reaction involves the use of a strongly alkaline reagent and care is essential in the adjustment of the balance of the reagents. This reagent also restricts the choice of sugar method that can be used on the deproteinised sample, and the conditions for the production of clear filtrates are sometimes elusive.

Somogyi [101] made an extensive study of deproteinisation reagents in connection with the measurement of blood sugar. In this he came to the conclusion that a zinc hydroxide precipitate formed in the reaction between zinc sulphate and sodium hydroxide provides a satisfactory method of deproteinisation for blood, giving a clear neutral filtrate that could be used in many of the sugar methods available at that time. This method has been used successfully for the deproteinisation of milk samples and both reducing and colorimetric methods for lactose can be used on filtrates prepared in this way.

The procedure is as follows. The two reagents (10 % (w/w) zinc sulphate in water and 0·5N NaOH) are prepared and slowly titrated against one another using phenolphthalein as indicator; the concentration of the 0·5N NaOH is then adjusted so that equal volumes of the reagents are exactly neutral.

The reagents are added to the sample diluted in a suitable volume of water in a volumetric flask. The volume of each reagent should be $\frac{1}{10}$ of the final volume. After addition of the two reagents, the mixture is adjusted to volume and mixed, and after about 15 min the solution is filtered.

Reducing Substances

If reducing methods are to be used, then non-sugar reducing substances must be removed. Treatment with lead acetate removes many potential interfering reducing substances but some reducing amino acids are not precipitated by this procedure. Treatment with ion-exchange resins is probably the best approach and, where ionic reducing substances are suspected, the sugar extract should be treated with suitable resins and the results before and after such treatment compared.

THE EXTRACTION OF SUGARS

Extraction with Aqueous Alcohols

The extreme solubility of the free sugars in aqueous solutions makes these the obvious choice as extracting media. There are several disadvantages, however, in the use of aqueous solutions for extracting the free sugars from foods. The first is that the aqueous reagents extract many substances that interfere with the subsequent measurement of sugars, and so quite extensive deproteinisation and decolorisation procedures are frequently necessary before measurements can be made. There are also technical problems in the extraction of some foods with aqueous solutions, and filtration of these extracts is frequently difficult and time-consuming.

Most analytical approaches to the extraction of the free sugars have therefore tried to combine the twin requirements of complete extraction of sugars and minimal extraction of interfering substances, in procedures that yield clear filtrates with the minimum of manipulation. However, the use of more specific techniques for sugars, for example linked biochemical reactions, may well be carried out quite satisfactorily on simple aqueous extracts and the analyst dealing with an unfamiliar material should probably try an aqueous extraction first; using a hot aqueous extraction he can at least be confident that he will get complete extraction!

The free sugars are generally soluble to a significant extent in aqueous alcoholic solutions and, as proteins and virtually all polysaccharides are insoluble at alcoholic strengths above 70–75 % (v/v), these reagents have formed the basis of most extraction procedures in unified analytical methods.

Several different techniques for extraction with aqueous alcohol have been employed and ethanol, methanol, and iso-propanol have been used quite extensively. The disaccharides, particularly lactose, are rather sparingly soluble in alcohol and in some cases it has been customary to extract at around 50 % (v/v) ethanol and subsequently increase the alcoholic strength to precipitate polysaccharides. In this way relatively high concentrations of many disaccharides can be brought into solution. A general precaution applies to all methods of extraction; if one wishes to determine the proportions of oligosaccharides in a foodstuff, then the extracting medium must be neutral and remain neutral during the extraction. Significant amounts of organic acids are present in many fruits and these will bring about a partial hydrolysis unless they are neutralised. Solid calcium carbonate is often added to the extraction medium but a more sophisticated approach using a buffered alcoholic extractant is preferable. A similar *caveat* applies where readily hydrolysable polysaccharides are present, as for example the fructans of cereals and many plants and the labile arabinofuranosyl side chains of some structural polysaccharides and gums.

Extraction in a Soxhlet-type Apparatus

Continuous extraction in a Soxhlet apparatus has been used by many authors. In most instances 80 % (v/v) ethanol has been the preferred extractant [74]. A few hours (6–8) seem to be sufficient to extract the free sugars from most plant products and probably most foods. The sample should ideally be fairly finely divided, although coarsely chopped samples have been extracted in this way. There are a few objections to this type of method but some of these are academic for the analyst. The first is that unless the sample has been dried the initial extraction is at less than 80 % (v/v) alcohol. This is true but may be a reason why complete extraction is obtained; 80 % (v/v) ethanol is not an azeotropic mixture and the actual concentration of ethanol in the extraction apparatus shows cyclical changes as the extraction proceeds. This would not greatly affect the overall efficiency of extraction but may make the process slower than it would otherwise be.

A more serious objection is that the extract is heated for some considerable time. The extracts prepared by this method do tend to darken slightly as the extraction proceeds; it is, however, difficult to detect much destruction of sugars and as pigments and traces of lipids are also extracted this progressive darkening may be inconsequential. However, a general principle of importance to the analyst is that, if possible, one should avoid prolonged heating of sugar solutions.

Extraction at a Lower Alcohol Strength

This forms the basis of several AOAC procedures for the extraction of sugars, particularly from cereal foods [4].

The sample is suspended in a volumetric flask half-filled with 50 % (v/v) ethanol and the flask is heated in a boiling water-bath for one hour, with a filter-funnel in the neck of the flask to minimise evaporation. At intervals during the process the contents of the flask are mixed by swirling. After cooling, the flask is made to volume with either absolute or 95 % (v/v) ethanol, the contents mixed and allowed to stand. The proteins and polysaccharides settle quite rapidly and, after leaving overnight, it is usually possible to decant the clear supernatant or to withdraw an aliquot with a pipette for analysis. This method will bring large amounts of free sugars into solution and provides a very mild treatment of the extract; it is also easy to incorporate calcium carbonate in the extracting flask to neutralise any acidity.

Two technical points are worthy of attention. First, the alcoholic solution has a high coefficient of thermal expansion and careful temperature control during adjustment of volumes and removal of extract is essential in accurate work. Second, there is a small volume correction to be made for the residue.

This can be avoided if the contents of the flask are filtered, for example through a Buchner funnel, and the filtrate made to volume.

Extraction with Methanol
Extraction with hot 85 % (v/v) methanol either by refluxing the sample with the solvent, or by repeated extraction of a small sample (5 g) in the solvent just brought to boiling point, has also been used to prepare extracts of foods [104]. Hot methanol at this strength also has a considerable capacity as a lipid solvent and occasionally lipid will separate from the extract on cooling. Filtration is usually rapid.

Extraction with iso-propanol
Friedmann and his colleagues [44] used two different procedures for extracting free sugars; 20 % (v/v) ethanol and 40 % (v/v) iso-propanol. They found that in heat-treated foods the dextrinised starch dissolved in the ethanol and in preference they used iso-propanol, the extraction being made at room temperature.

Treatment of Alcoholic Extracts
The alcoholic extracts contain, in addition to sugars, some lipids, pigments and the free amino and organic acids present in the sample. The next stage in the preparation of the extract must therefore be to remove interfering substances and to produce a solution of sugars suitable for the methods to be used subsequently.

Most sugar methods are sensitive to high concentrations of alcohol and, for the reducing sugar methods in particular, alcohol must not be present in the extract. The extract may also need to be concentrated for analysis, although in practice this is rare; the need for dilution is much more common.

Slow evaporation of the extract, preferably in a rotary evaporator under reduced pressure, is the most satisfactory way of both concentrating the extracts and removing alcohol. Gentle heating on a water-bath in a well ventilated fume-hood can be used if a rotary evaporator is not available but frequent attention is necessary to avoid the sugar solution drying out on the walls of the vessel (a tall-form beaker is very suitable) by rinsing the walls with water at intervals.

Before any concentration is started the pH of the extract should be adjusted to neutral with small amounts of sodium carbonate.

The concentrated extract should then be treated with a small amount of lead acetate (as a saturated solution) (0·12–0·2 ml) and made to a suitable volume. This will precipitate most pigments and produce a neutral clear solution. Reducing amino acids may still be present but, with most foods, treatment of an extract prepared in this way with ion-exchange resins does not usually result in any pronounced effect on the reducing sugar values.

Although removal of the alcohol is obligatory for many sugar methods, some colorimetric procedures can tolerate the presence of modest amounts of alcohol. The reactions with anthrone and orcinol are not affected by alcohol concentrations of the order of 5 % (v/v), and the sensitivity of these reactions is so high that it is frequently possible to dilute the alcoholic extracts accordingly. For accurate work, however, standard solutions containing equivalent concentrations of alcohol should be used to construct the standard curves in these cases.

MEASUREMENT OF SINGLE SUGARS

Where a single sugar is present in the extract or where a value is required to be expressed according to an established convention (as for example reducing sugars calculated as glucose), the analyst can use a relatively non-specific method.

A more specific procedure might be preferred, however, if it conferred some advantage, such as the saving of time, improved precision, or where it reduced the amount of preparatory work to be done on the extract.

In most instances the analyst would rely on a reducing sugar method, provided of course that he had access to the appropriate calibration for the sugar being measured. In general, non-specific procedures, such as a reducing sugar method, should only be applied where one knows, either by qualitative examination or by experience, the composition of the sugars in the extract.

Reducing sugar methods can be applied to measure free reducing sugars in the presence of the non-reducing sucrose and such methods form the basis for measuring sucrose by determining the increase in reduction after hydrolysis of the sucrose. These procedures depend for their accuracy on a semi-quantitative knowledge of the composition of the solution being analysed.

Reducing Sugar Methods

Copper Reduction Methods

A wide range of methods has been used for the measurement of reducing sugars. Many variations of the original Fehling's reagent have been used, which have special merits for particular foodstuffs. The principal aim of most of these modifications has been to improve the precision of this reduction method and to remove the cause of the major variations in the yield of cuprous oxide. Of these, the alkalinity of the reagent, the rate and time of heating and the concentration of the sugar in the samples appear to be most important. Several techniques have also been used for measuring the cuprous oxide formed. A number of unified procedures where the conditions have been exhaustively studied and where detailed calibration data have been obtained are available and two of these, the Lane–Eynon method and the Munson and Walker method, are the most regularly used. They are the standard reducing sugar methods recommended by the ICUMSA committee [128].

In the Lane–Eynon method (*see* Chapter 8 (Reducing sugar methods) for detailed procedure), the sugar solution is titrated against the hot Fehling's reagent in two stages; the first when the bulk of the solution required to effect almost complete reduction is added, and the second dropwise to an end-point with methylene blue. Careful control of the heating is required and for the most accurate analyses two titrations are necessary; the first to establish the approximate volume of the test solution to effect reduction, and the second to measure the precise volume required. This method is very widely used and gives satisfactory results in practised hands.

In the Munson and Walker procedure, a fixed volume of solution is heated under standard conditions with the Fehling's reagent and the cuprous oxide formed is filtered off and either weighed directly, measured volumetrically by titration with permanganate or thiosulphate, or measured electrolytically.

Again the method has been extensively studied and detailed calibration tables are available for a range of sugars. The detailed procedure is described in Chapter 8 (Reducing sugar methods).

Microcolorimetric or volumetric versions of copper reduction methods are also available, and these have formed the basis of some automated methods for the measurement of total reducing sugars.

Ferricyanide Reduction
The reaction takes place in alkaline solution and standardisation of the heating conditions is essential—the reduction can be measured either by iodimetric procedures or colorimetrically. The stability of the alkaline ferricyanide reagent and the fact that the experimental conditions are apparently less demanding than the copper reduction methods has made this procedure a method of choice in many automated reducing sugar methods. The version described in Chapter 8 (Reducing sugar methods) is a manual one but the essential features of the automated procedures can easily be derived from this basic method. Technicon Ltd have published Autoanalyser methods using the ferricyanide procedure [26].

Reduction of Tetrazolium Compounds [29]
This appears to be likely to provide a reducing sugar method of potential importance in the automation of sugar analysis and a manual procedure is described in Chapter 8 (Reducing sugar methods).

Colorimetric Condensation Reactions
The wide range of condensation reactions in strong acid solution leading to coloured products of potential use to the analyst was mentioned earlier in Chapter 2 (Chemical properties). The methods were reviewed by Dische in 1955 [35] and, since then, the general tendency has been to restrict the use of many reagents and to concentrate on the reagents that seem to be most suitable as analytical tools [130]. The methods described in Chapter 8 (Condensation reactions) fall within this category and in the main have been well-tried and tested.

The methods are not in themselves highly specific, although the conditions under which the reactions take place can be modified to increase the colour yield of one class of monosaccharides relative to another, and so improve specificity. When a single sugar species is present in an extract, this lack of specificity is of no significance but the analyst must be sure of the qualitative composition of the extracts before reporting the results obtained by a condensation reaction in terms of a single monosaccharide.

The methods described in Chapter 8 (Condensation reactions) comprise the phenol sulphuric method [130], which is a general method for all carbohydrate species; the anthrone procedure, which under the conditions described is reasonably specific for hexoses [96]; the ferric-chloride/orcinol/HCl method for pentoses [1]; a resorcinol method for ketoses [65] and a carbazole/borate/H_2SO_4 method for uronic acids [10].

The methods using phenol and anthrone are carried out at an acid strength that will hydrolyse oligosaccharides and polysaccharides; it is therefore important to avoid chance contamination with cellulosic dusts or fibres from filter papers. The capacity of the anthrone reagent to react with both monosaccharides and oligosaccharides makes the anthrone reaction of great value for total hexoses. The colour yield of galactose is only 0·54 that of glucose, and where lactose is being measured appropriate lactose standards should be used.

Technique. The general technique for the condensation reactions is very similar for all methods. The reaction is conveniently carried out in glass-stoppered tubes; the small amounts of organic solvents may occasionally dislodge a stopper when the tubes are placed in a heating bath but there is little danger of explosion due to the generation of excessive pressures.

The strong acid reagents are most conveniently dispensed using automatic pipettes but these have to be used with care with the more viscous sulphuric acid reagents.

Heating and cooling must be closely controlled and all the tubes in a series (blanks, standards and samples) should be placed in the water-bath or cooling bath simultaneously using a wire basket, or a rack with handles. The water-baths used must also be heated so that the immersion of the tubes does not cool the bath too much. If standards and samples are treated identically this should not produce too many problems. The water-bath must contain sufficient water to immerse the tubes above the level of the mixture in them.

The measurement of the absorbance can be made in any spectrophoto-meter at the appropriate wavelength. Some problems with bubble formation in the sulphuric acid reagents can occur at the measurement stage; in practice this usually means that the contents of the tubes were not adequately mixed before heating commenced.

Although with a very rigid technique many of the methods described will give reasonably reproducible results for standards run on different occasions, it is generally advisable to construct a standard curve with each series of samples.

Blank readings with some of these reagents tend to increase with the age of the reagent and, with the anthrone reagent especially, can be responsible for reduced precision. The recrystallisation of anthrone from ethanol usually reduces blanks with this reagent and the same is true for some batches of orcinol and carbazole.

Enzymatic: Biochemical Procedures
These provide very specific procedures for some monosaccharides and disaccharides and the improvements in enzyme technology over the last ten years have meant that enzyme preparations of high specificity and activity have become available commercially at prices which make them useful in routine analysis.

The available methods can be seen to fall into two classes. The first involves the use of specific enzymes to react with a sugar to produce a compound that reacts with a chromogen which is then measured. The glucose oxidase procedure is a method of this sort [11].

The second class involves the coupling to a biochemical reaction and following the changes in the reduction of nucleotides. These methods may also include preliminary hydrolysis of an oligosaccharide with a specific carbohydrolase.

In the clinical measurement of glucose, the glucose oxidase methods have virtually displaced the older reductiometric methods because of their specificity and their ease of operation. Their use in food chemistry is increasing and they should eventually provide the method of choice where glucose must be measured.

The main requirements for the use of enzymatic methods are a neutral or acid extract free from alcohols and heavy metals. The use of lead salts for cleaning up extracts for enzymatic methods is therefore forbidden, and alcohol must be removed by evaporation. As the application of these enzymatic methods develops, it may well be shown that these clarification procedures are unnecessary. Proteins are usually undesirable contaminants of the extracts. Deproteinisation with perchloric acid is suggested as a suitable method for use in conjunction with many procedures.

The purity and specificity of the enzymes are the limiting features in most of these biochemical procedures and the analyst must therefore depend very greatly on the integrity and quality control of the supplier. The use of careful control standards at intervals, and certainly when a new batch of reagents (enzymes or cofactors) are being used for the first time is essential if consistently reliable results are to be obtained.

THE ANALYSIS OF MIXTURES

Extracts of foods that contain a single species of sugar are relatively rare in practice. The measurement of the free sugars in most foodstuffs therefore involves the analysis of mixtures.

The correct approach to this type of analysis depends on a number of factors. Of these, two are of primary importance. First, the qualitative composition of the mixture, and second the actual requirements of the analysis.

Many analysts are concerned with measuring carbohydrates in a limited range of foods where the composition of the free sugars is relatively constant and where for many purposes, for example quality control, total reducing sugar values expressed as invert sugar or glucose may be perfectly adequate.

In other situations a complete analysis of the different carbohydrate species is desirable.

There are a number of different options available to the analyst, ranging from the use of a non-specific reducing sugar (or condensation reaction) method to the separation and analysis of the individual components.

Qualitative Examination of Sugar Extracts

The correct interpretation of the results obtained from the application of any sugar method to an extract from a foodstuff depends on a knowledge of the qualitative composition of the extract.

The knowledge may be based on past experience with the food in question, 'received knowledge' in one sense, or it may be based on a routine qualitative stage in the analysis.

Where a range of foodstuffs is being examined, and particularly when processed foods are included in the samples, it is generally wise to have a set routine for qualitative examination of all extracts. Either thin-layer or paper chromatography provide the analyst with suitable procedures, and the choice is a matter of personal preference.

Paper Chromatographic Examination

The paper chromatographic analysis of sugar mixtures has been reviewed on many occasions and a number of authoritative reviews are available [55,

66, 69]. Some general observations about the choice of technique may however be useful.

Solvents. Of the many solvent systems that have been used for the separation of sugar mixtures, two basic series have evolved; those based on 1-butanol and those based on ethyl acetate. Butanol-based mixtures tend to give lower mobilities than the ethyl acetate mixtures, although these tend to deteriorate more rapidly on storage. The food analyst will probably find it better to use one or two particular solvent systems which give reasonable resolution of the sugars commonly found in foods, rather than be constantly ringing the changes from one system to another (Chapter 8 (Qualitative procedures)).

Technique. Whatman no. 1 (or its equivalent) has become accepted as the paper of choice for sugar separation and most published tables of mobilities concern this paper.

The descending method also appears to offer the best separation and, when using ethyl acetate mixtures, it is usually convenient to let the solvent run off the paper to give improved separation.

In general, the approach to the application of the sample should be a quantitative one and care should be taken with the levelling of the tank and the dimensions and configuration of the paper, so that the day-to-day running of the system becomes a routine and semi-quantitative information can be gleaned from the chromatograms.

Detecting sprays. Here again a battery of reagents are available. The analyst will probably find it best to standardise on one or two, and use these on a routine basis. Many authors feel that ammoniacal silver nitrate still provides the best detecting agent, but my own preferences are for spray reagents that give qualitative information about sugar types on the chromatogram. For this reason I prefer aniline hydrogen phthalate or naphthoresorcinol (Chapter 8 (Qualitative procedures)).

Thin-layer Chromatography
The remarks on paper chromatography also apply to this technique, and there are several extensive reviews [110]. The general consensus of opinion and practice is that it is advisable to use a silica gel phase that has been impregnated with borate if good separations are to be obtained, and it is extremely important to avoid overloading the plates. For routine work the analyst will probably find that the purchase of prepared plates is extremely useful both from the point of view of the time saved and in consistency of the results from day to day.

A greater variety of reagents are suitable for thin-layer than with paper because of the resistance of the plate to strong acid. Naphthoresorcinol in phosphoric acid seems to provide a particularly useful reagent (Chapter 8 (Qualitative procedures)).

Analysis of Mixtures—Without Separation
The most frequently encountered mixture is one containing an equimolar mixture of glucose and fructose derived from the hydrolysis of sucrose, invert sugar, and sucrose itself in varying proportions.

Chemical Approaches
Mixtures of this type have been the concern of the analysts in the sugar industries for many years and the behaviour of the copper reducing method towards these mixtures has been studied extensively.

When mixtures of this type are being analysed, the copper reducing methods (Chapter 8 (Reducing sugar methods)) will give satisfactory results by using the appropriate columns of the reference tables.

These tables give values for the reducing equivalents of invert sugar alone and in the presence of some sucrose. Where the ratio of glucose to fructose is not unity, the error introduced by interpolation based on the actual glucose to fructose ratio, which can be assessed from inspection of paper or thin-layer chromatograms, is small. The effect of the sucrose concentration on the reduction of invert sugar mixtures is quite significant but an approximation for the sucrose content is usually adequate.

The analysis of invert sugar and sucrose mixtures by the application of reduction methods before and after inversion of the sucrose is a standardised and reproducible technique. Inversion of the sucrose can be carried out enzymatically or by treatment with acid at room temperature. The AOAC procedure uses hydrochloric acid at about $1.2N$, overnight at a temperature greater than $20\,°C$. The acid must be neutralised before the measurements of reducing sugars are made.

The presence of reducing disaccharides, for example lactose or maltose, in a mixture makes the use of a single reducing sugar determination of little value. These disaccharides are not appreciably hydrolysed in conditions that bring about the hydrolysis of sucrose, so that they can be measured as the monosaccharides after a more vigorous hydrolysis. Hydrolysis in $0.4N$ H_2SO_4 for one hour at $100\,°C$ will convert these disaccharides completely to the monosaccharides with only minimal destruction of the fructose in any sucrose present. However, these techniques must be regarded as approximations and in many ways obsolete by comparison with more delicate biochemical methods to be discussed later.

Differences between the reduction equivalents of glucose and fructose and other reducing sugars in different methods, for example copper reduction methods and ferricyanide reduction, have provided ingenious ways for the analysis of mixtures [133]. In these the reduction values are measured by the two methods and used as simultaneous equations to solve for the two components. McCance *et al.* [74] used this approach for measuring the proportions of glucose and fructose in fruits and vegetables and the results obtained agree well with more recent methods. However, this approach must be seen as an approximation, using the best methods available at the time.

Similar remarks apply to the use of microbiological techniques for the assay of mixtures of sugars.

The use of more specific condensation reactions can provide useful information about the proportions of the major classes of sugars present. The limitations of this approach lie in the lack of genuine specificity of the reagents. Corrections for cross interference can be applied, however, and these methods can provide values for hexoses, pentoses and uronic acids in a mixture with reasonable reliability [104].

The use of controlled mixing of sample solution and acid and of heating conditions in the analysis of sugar mixtures has been examined for the anthrone reaction but in practice these are extremely difficult to control on a routine basis and such methods must be regarded with a little caution.

Biochemical Approaches
The biochemical procedures based on coupled enzyme reactions in which the rate of production of reduced nucleotides is followed provide the most delicate and specific methods for the analysis of sugar mixtures. Procedures for the measurement of free glucose and fructose are available and for the measurement of sucrose, maltose and lactose after specific enzymatic hydrolysis. The limitations of the procedure depend on the specificity of the enzyme preparations and on the ability of the analyst to produce an extract of the food containing the sugars which can be introduced into the enzymatic procedures. Fortunately, this does not seem to involve great difficulties (Chapter 8 (Biochemical procedures)).

Analysis of Mixtures—with Separation
The general lack of real specificity in the analytical methods for sugars, with the exception of the biochemical procedures, implies that accurate analysis of mixtures must be preceded by separation. In the quantitative analysis of

sugar mixtures, three approaches have been made based on quantitative paper chromatography, liquid chromatography on ion-exchange resins and gas–liquid chromatography.

Of these, paper chromatography must be regarded as semi-quantitative, although in many workers' hands quite good levels of precision and accuracy have been obtained and useful results reported.

Sugars are easily eluted from paper and, apart from problems caused by interference from cellulose fibres, the micro-reduction or condensation methods have been applied with success.

Ion-Exchange Chromatography
The main problem with sugar chromatography is the very low degree of ionisation of carbohydrates in general and in most cases ion-exchange separation has only been achieved when using the borate complexes of the sugars. However, this has proved a reasonably reliable method for separating sugar mixtures and a number of elution systems have been suggested [60, 63, 83, 125]. In the early work on the ion-exchange separation of sugars, aqueous ethanolic buffers were used. The use of volatile buffers produced some technical problems in achieving the best separation as the resin columns have to be maintained at elevated temperatures. Difficulties were also experienced with the colorimetric assay of the eluates and with the life of the manifolds in automated analytical systems. In the method [41] described in Chapter 8 many of these technical problems have been resolved and a chromatographic system has been developed that is suitable for the analysis of the sugars likely to be found in extracts of foods.

The principal technical problems connected with automated sugar analysis are centred around the choice of the colorimetric method used in analysing the column eluates. Most authors have used a condensation reaction, but this has usually resulted in problems with base-line stability and in the life of the tubing used to pump the strong acid reagent. In general, the automated systems must be constructed so that the maximum of glass tubing (and the minimum of plastic) is used. The pump tubes for the acid reagent must be of Acidflex tubing (Technicon), although if ethanol-free buffers are used, normal pump tubing can be used for the sample.

In general, ion-exchange chromatography is still a fairly time-consuming procedure compared with, say, gas–liquid chromatographic separations. However, the system does not involve the lengthy stages necessary for the preparation of suitable derivatives and the overall productivity of the ion-exchange chromatographic separation and analysis may be comparable with gas–liquid chromatographic analysis.

High-pressure chromatography may be expected to provide a powerful tool in the separation and analysis of sugar mixtures in the near future.

Gas–Liquid Chromatography

The development of this method of separating mixtures of organic compounds has been extremely rapid and gas–liquid chromatography as an analytical tool has had a profound effect on virtually every field of analysis. This has also been true in the carbohydrate field. The main technical obstacle to the use of this technique for carbohydrates was the choice of suitable volatile derivatives, as the carbohydrates themselves were not sufficiently volatile to be analysed directly [15, 37].

The fully methylated derivatives were soon shown to be suitable and the separation of these derivatives with the gas chromatograph in combination with a mass spectrograph has proved an exceptionally powerful tool in the structural analyses of polysaccharides.

Trimethylsilyl derivatives were shown by Sweeley and his colleagues [112] to be suitable derivatives for the quantitative analysis of sugar mixtures. A wide range of silylating reagents is now available and these derivatives can be prepared rapidly and easily (Chapter 8 (Gas–liquid chromatography)) [90]. The derivatives are quite stable but are readily decomposed to give the free sugar. The major technical problem associated with the preparation of these derivatives arises from the sensitivity of the silylating reagents to moisture, which necessitates the reaction being carried out under anhydrous conditions, usually in pyridine.

Trimethylsilylation also produces derivatives corresponding to the equilibrium mixture of the individual sugars, so that the four possible derivatives are often formed. This means that chromatograms of sugar mixtures are usually rather complex. This can be a disadvantage if the method is being used as a routine analytical tool, although if one type of mixture is being analysed regularly this disadvantage is minimal.

Mixtures of the trimethylsilyl derivatives of the monosaccharides and the lower oligosaccharides can be separated readily on instruments that have the facility for temperature programming of the column. Separation of the derivatives of lactose and sucrose is not usually complete, however, but mixtures containing these sugars can be analysed by carrying out the trimethylsilylation and chromatography before and after hydrolysis of the sucrose with invertase.

The problems associated with resolution of the trimethylsilyl derivatives of the anomeric forms have led to a considerable amount of work undertaken to simplify the chromatograms.

Reduction of sugars to the acyclic alditols provides a way of avoiding difficulties associated with the analysis of anomeric forms. The alditols are normally prepared by reduction with borohydride; a reaction that takes place under mild conditions with reasonable speed. The acetyl derivatives of the alditols have proved to be the most suitable ones for gas–liquid chromatographic systems [29]. Analyses of the cell wall polysaccharides using gas–liquid chromatographic separations and analysis of the alditol acetates have provided a considerable insight into the fine structure of the cell wall [3].

The alditols appear to form complexes with boric acid which inhibit the acetylation reaction and it is essential to remove excess borate before attempting to acetylate the mixture. Evaporation with methanol and the removal of the borate as methyl borate has been found to provide a mixture that can be acetylated quantitatively (Chapter 8 (Gas–liquid chromatography)).

Satisfactory separation of the alditol acetates can be achieved with a column running under isothermal conditions.

The conditions used in the analysis of sugar mixtures with both trimethylsilyl derivatives and the alditol acetates, together with a range of other derivatives, have been collected together in a review by Dutton [37], and the methods described in Chapter 8 (Gas–liquid chromatography) are some of the many that have been found useful in the quantitative analyses of mixtures of the type found in food analysis.

Quantitation invariably requires the use of internal standards, which ideally should enable the quantitation of the whole process of the preparation of derivatives. It is difficult to suggest suitable substances that fit this criterion precisely. 2-Deoxy-glucose has been used in separations of the trimethylsilyl derivatives and myo-inositol in the separation of the alditol acetates, although this latter compound does not permit any assessment of the efficiency of the reduction stage.

As a routine method in food analysis gas–liquid chromatography provides a method of power but the preparation of derivatives is somewhat time-consuming.

SUMMARY

The Choice of Methods for the Analysis of Sugars in Foods

The choice of methods in the analysis must be based on a knowledge of the composition of the mixture of sugars in the food in question. This can be made by a 'key' type of approach.

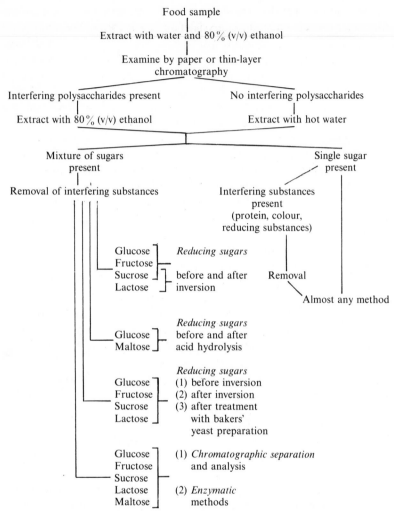

FIG. 3.2.   Scheme for the choice of method for the analysis of sugars in foods.

In the analysis of an 'unknown' food the first stage should be a qualitative (or better a semi-quantitative) examination of aqueous and aqueous ethanolic extracts.

If a single sugar is present and there are no interfering substances almost any method of analysis can be used. The presence of interfering substances means that some method of removing these substances is usually necessary;

here the appropriate treatment depends on the method to be used, for example, an enzymatic method may require the minimum of treatment, such as deproteinisation, whereas a reducing sugar method will require the removal of non-sugar reducing substances and other substances likely to react with the Soxhlet reagents.

The choice of method when a mixture of sugars is present is more difficult. Glucose, fructose and sucrose, and lactose and sucrose can often be analysed quite well with reducing sugar methods.

The presence of maltose and its higher homologues can usually be tackled by measuring the increase in reducing sugars on dilute acid hydrolysis ($0.4N$ H_2SO_4 for 1 h at 100 °C).

Lactose provides an additional complication but simplified procedures can be devised based on the removal of fermentable sugars with bakers yeast; however, the use of specific enzymes provides the better approach.

Analysis of the complex mixture of glucose, fructose, sucrose, lactose and maltose which occurs in many human mixed diets is difficult using reducing sugar methods alone. A combination of enzymatic and acid hydrolysis can be used but the most useful approach is either chromatographic separation and analysis or the specific enzyme methods. Quantitative paper chromatography can be used but ion-exchange chromatographic separation is preferable. The trimethylsilyl derivatives from this mixture present a very complex chromatographic trace, and sucrose is not resolved completely from lactose. The alditol acetate derivatives cannot be used, because glucose and fructose give the same alditols.

The use of certain enzymes provides a delicate and highly specific method for the analyses of the mixtures of sugars frequently encountered in food analysis. Detailed studies using these methods have been made of the sugars in fruits and vegetables.

The principles involved in the choice of method are summarised in Fig. 3.2.

CHAPTER 4

The Measurement of Starch, its Degradation Products and Modified Starches

FORMS IN FOODS

This chapter is concerned with the measurement of starch in foods. In the raw food this polysaccharide is located in discrete granules, which have structures characteristic of the foodstuff in question [31, 50]. The visual appearance of the starch granules can be used for the qualitative identification of the foods present in a mixture [58].

The starch granule has been the subject of a considerable amount of work, and the arrangement of the carbohydrate molecules within the granule itself is highly ordered [78]. The granules appear to have a lipoprotein coat and many isolated starches have a definite waxy texture.

On heating in moist conditions the starch absorbs water and swells, leading to the eventual rupturing of the granules and the extrusion of the water-soluble (or strictly dispersible) starch. The analyst dealing with foods that have been cooked will therefore be dealing with samples where many, if not all, of the starch granules have been disrupted in this way.

Dry heating of starch leads to the production of dextrins, a reaction that initially involves some depolymerisation and oxidation of the molecule. Continued heating leads to the progressive degradation of the molecule, and to the formation of gums where profound structural modifications of the initial molecule have occurred. In many baked products some starch has been so degraded that the molecular species present range from the original starch molecules to carbon.

In addition to the heat degradation products of starch, found to a greater or lesser extent in all heated foods, many foods also contain acid or enzymatically degraded starches. These 'glucose syrups' are widely used in the food processing industry and in the production of confectionery [14]. The molecular species present in these syrups varies depending on the method of manufacture but, in general, a range of molecular sizes is usually found. The mixtures include some free glucose, maltose, maltotriose and maltotetrose, and these will usually be considered as falling within the *free sugar* fraction. However, for analytical purposes many of the procedures for the measurement of starch can be used for these products and it is convenient to consider them in this chapter.

Starch can also be chemically modified to introduce various kinds of cross-links; these modified starches possess a number of properties that are of importance in food technology [56]. For example, they will form cold setting gels or they will withstand a freeze-thaw cycle without retrogradation (a property of the amylose moiety of starch that leads to precipitation). These modified starches are therefore important to the food analyst, although the concentrations of them in most foods are usually low.

CHEMICAL STRUCTURE

The main features of the chemical structure of starch were outlined in Chapter 2, p. 12.

Most if not all food starches appear to contain two types of molecule, amyloses with molecular weights of the order of 10^3 to 5×10^5. Amylose is essentially a linear molecule in which the glucose residues are linked 1,4α. The configuration of the molecule appears to be helical with a relatively large internal core to the molecule. Amylose disperses readily in water but the chains readily recombine and reprecipitate in the process known as retrogradation. Amylose forms complexes with iodine and is responsible for the intense characteristic blue colour given by starch with iodine in potassium iodide. The iodine complex is of the inclusion type, and the tertiary structure of amylose is such that these complexes form very readily.

The other type of molecules found in starch are the amylopectins; these have much higher molecular weights than the amyloses (up to 10^6) and are much less dispersible in water. The molecule appears to have a highly branched structure with 1,4 and 1,6α glucosidic linkages. The 1,6 linkages appear to occur at intervals of about 25 glucose residues, and the molecule appears to be randomly branched.

The amylopectin molecule does not appear to form stable complexes with iodine, but gives a very pale red colour in its presence.

The proportion of amylose to amylopectin in a starch has important effects on the physical properties of the starch and affects the suitability or otherwise of a food for some technological processes. Table 4.1 gives some typical values for the proportion of amylose in some starches.

The measurement of the proportion of amylose in a starch preparation usually involves the formation of the iodine complex and potentiometric

TABLE 4.1
AMYLOSE CONTENT OF SOME STARCHES [50]

Food starch	Amylose (g/100 g)
Barley	22
Oat	27
Wheat	26
Maize	28
Broad bean	24
Pea (smooth-seeded)	35
(wrinkled-seeded)	66
Parsnip	11
Potato	40
Apple	19
Banana	16

titration of the iodine. The principles of this procedure are described by Greenwood [50].

MAJOR OPTIONS IN THE ANALYSIS OF STARCH

Most analytical procedures for the measurement of starch involve hydrolysis of the polysaccharide and the measurement of the products of hydrolysis (usually as glucose). Quantitative measurement of starch using the reaction with iodine has also formed the basis of some methods. Polarimetric procedures for starch have also been developed for cereal foods and these will be discussed in detail below.

Direct Acid Hydrolysis

Dilute acid hydrolysis of starch usually gives a near theoretical yield of glucose and this simple procedure has formed the basis of some analytical methods. The strength of acid used for hydrolysis may be varied but 0·4 N H_2SO_4 will convert starch quantitatively to glucose after 4 h under reflux: the use of stronger acids is unnecessary. Most foods containing protein, and to a lesser extent fat, form condensation products between amino acids and carbohydrates when subjected to acid hydrolysis, which limits the usefulness of this direct approach. The progressive darkening of acid hydrolysates as heating is continued is also indicative that destruction is taking place, although it is frequently difficult to detect using reducing sugar methods.

Darkening of the hydrolysates and the formation of 'humin' can be minimised if a high ratio of hydrolysing acid to sample is employed. Most cereal foods give clear colourless hydrolysates when hydrolysed for 4 h under reflux in 0·4N H_2SO_4 at concentrations of between 100 and 200 mg/litre.

Free sugars should normally be extracted from the sample before acid hydrolysis is started. This is essential if sucrose is present, as the fructose formed on hydrolysis decomposes quite rapidly under acid conditions and a true value for the increase in sugars after hydrolysis cannot be obtained.

A major objection to direct acid hydrolysis for the measurement of starch is that all other acid-hydrolysable polysaccharides will also be measured. These will include the majority of the non-cellulosic polysaccharides (the pectins, hemicelluloses, etc.), which will produce reducing sugars. The use of a glucose-specific method such as the glucose oxidase method (Chapter 8 (Biochemical procedures)) or the glucose hexokinase method will remove some of these objections. However, many cereals contain β-glucans (with 1,3 and 1,4β glucosidic bonds) and these will yield glucose, resulting in an elevation of the starch values. In many instances, however, direct acid hydrolysis followed by the measurement of glucose, preferably by an enzymatic method, provides a satisfactory way of measuring starch (*see* Chapter 8 (Starch by direct acid hydrolysis)).

In other cases, however, the presence of interfering polysaccharides or other compounds will make an alternative approach essential. In these circumstances two approaches are available, selective extraction or selective hydrolysis.

Selective Extraction

The two major types of molecule found in starch differ appreciably in their solubilities in aqueous solvents. For example, hot water extraction of the sample will remove a considerable proportion of the amylose and the dextrins, leaving the bulk of the amylopectin in the residue.

The susceptibility of the dextrins to dissolve in water is important when the analysis of cooked products is planned, as water extraction is recommended in some cases for the extraction of sugars prior to the measurement of starch. Dextrins are often appreciably soluble in aqueous alcohols of around 20 % (v/v) and caution is necessary when using these reagents for extracting samples.

Calcium Chloride

Extraction with $CaCl_2$ solution has been widely used in the analysis of cereal starches by polarimetric methods and appears to give reproducible results [43]; other polysaccharides are soluble in this reagent and some care is necessary in the application of this method to other types of foodstuff. It remains, however, a useful method for cereals (Chapter 8 (Selective extraction of starch)).

Perchloric Acid

Perchloric acid appears to give a reasonably selective extraction of starch from cereal-based mixtures and has been used to provide the basis for a method of measuring starch either colorimetrically by direct application of the anthrone reagent [21] or by precipitation of a complex with iodine [94] (Chapter 8 (Selective extraction of starch)).

Dimethyl Sulphoxide

Extraction with dimethyl sulphoxide, usually with the addition of acid, has also been used to bring starch into solution as the basis for a subsequent enzymatic hydrolysis.

In summary, however, most selective extraction procedures have been developed for specific classes of foodstuff and, while many of them may have a much wider field of application, this is by no means certain. The analyst must therefore take suitable precautions when using these extraction methods with materials that have not been tested.

Selective Hydrolysis

Selective hydrolysis of starch is virtually impossible using chemical procedures because many polysaccharides present in foods are considerably less resistant to hydrolysis than the starch molecules.

This hydrolysis must therefore be based on the use of enzymatic procedures, and a wide range of enzyme preparations has been and is recommended for this purpose. However, it is important to remember that the use of a particular enzyme preparation is, in effect, defining starch as the polysaccharide or group of polysaccharides hydrolysed by the enzyme in question.

The mixture of amylolytic enzymes produced by fungi was the first preparation to be used in starch analysis. The most important of these was Takadiastase, a mixture of α and β-amylases. The products obtained by the action of this enzyme on starch are a mixture of maltose and glucose. McCance *et al.* [74] when using this preparation had to employ two reduction methods and then solve the simultaneous equations to measure the composition of the mixture. The preparation of Takadiastase used by Southgate [103], however, was more efficient and gave a very nearly theoretical conversion of starch to glucose. McCrae [71] used another commercially available amylase mixture ('Agidex' BDH) in his method for measuring α linked polysaccharides. In general, however, these mixed fungal preparations could not be regarded as truly specific. Takadiastase preparations, for example, have been shown to have some activity towards proteins and even cellulose, although under the analytical conditions that were usually employed interferences from the hydrolysis of other polysaccharides was minimal and the slight proteolytic activity could even be considered advantageous.

Recently amyloglucosidases of greater specificity have become available and in many respects these have superseded the older enzyme preparations. The majority of commercial preparations are still of fungal origin but improved enzymatic purification procedures have enabled the manufacturers to produce very active and specific preparations. Details of the use of these preparations are given in Chapter 8 (Enzymatic hydrolysis of starch), but some general principles surrounding the use of these enzymes must be considered here.

Preparation of the Sample
Removal of sugars is not essential, as free glucose can easily be measured before the enzymatic hydrolysis, provided that interfering substances are not present. It is, however, possibly easier to work with a sample from which sugars have been removed.

Most enzymes will not attack intact starch granules and it is usually essential to 'gelatinise' the starch in the sample, either by heating it with water or by autoclaving the sample with water. Even foodstuffs that have been subjected to heating need to be treated in this way before enzymatic hydrolysis is started.

Hydrolytic Conditions
Most amyloglucosidase preparations have pH optima in the range of 4·5–5·5 and hydrolysis is usually carried out in acetate buffer. Table 4.2 gives some typical conditions for different preparations. Heavy metals and other substances likely to inactivate enzymes should not be present.

The preparations themselves have very high activities but it is usual to limit the amount of starch present and to provide a considerable excess of enzyme. If the starch in the sample can be dispersed into the buffer only a

TABLE 4.2
COMPARISON OF CONDITIONS FOR THE ENZYMATIC HYDROLYSIS OF STARCH WITH FOUR PREPARATIONS

	Conditions for hydrolysis				Method of measuring products	Reference
	pH	Buffer	Temperature (°C)	Time (h)		
Rhodosyme-s	4·7–4·8	0·04M acetate	50	6[a]	Ferricyanide	44
Agidex (BDH)	4·5	0·2M acetate	60	24	Glucose oxidase	71
Takadiastase	4·5	0·04M acetate	37	18	Reducing sugars Anthrone	103, 104
α-Amyloglucosidase	4·6	0·1M acetate	20–25	0·25[a]	Hexokinase	20

[a] Extraction method involves some initial acid hydrolysis.

short incubation time is necessary, but with other samples a more protracted incubation, up to 18 h, is necessary. In these cases it is usual to add a bacteriostatic agent such as toluene to the incubation mixture.

Measurement of Products
A glucose-specific procedure provides the most satisfactory method of measuring the products of hydrolysis, provided that the enzyme preparation is capable of bringing about complete hydrolysis. A method of this type is described in Chapter 8, p. 129. Where the products include maltose in addition to glucose, as occurs with some preparations, a less specific method such as the anthrone reaction is required to measure the products.

It is usually necessary to deproteinise the enzymatic hydrolysate before measuring the products of hydrolysis but with some specific methods this does not appear to be necessary.

Removal of Interfering Polysaccharides
Some water-soluble non-cellulosic polysaccharides are partially soluble in buffers used to disperse the starch and the process of gelatinisation may result in appreciable amounts of these polysaccharides being present in the hydrolysate. If a condensation reaction is being used in the measurement stage, it is necessary to precipitate and remove them before measurement.

Addition of ethanol to give a concentration of the order of 80 % (v/v) will result in the precipitation of most polysaccharides and has the additional value of removing the added enzyme at the same time. The analysis of the ethanolic solution then follows a similar procedure to that described for the analysis of free sugars (Chapter 3).

STARCH DEGRADATION PRODUCTS

The dextrins formed by the action of heat on the starch in foods will be measured with the starch by most of the methods considered above. The more extreme the heating, the greater the possibility that resistant linkages will be formed in the starch molecules. These will be of little consequence if an acid hydrolysis procedure is being used, although errors can arise if the reagent used to extract sugars also dissolves dextrins. Preliminary observations of enzymatic hydrolysis of starch in heated foods suggests that some resistant linkages are formed. The nature of these linkages is not clear but with foods that have been subjected to high temperatures, such as hard-baked products, the conventional methods of gelatinising the sample can result in a low value being reported for starch.

In these cases autoclaving and possibly a preliminary proteolytic treatment is required to get maximal yields with subsequent amyloglucosidase hydrolysis. It may well be that the proteolytic activity associated with some Takadiastase preparations was of unforeseen value.

The food analyst also has to consider the analysis of a range of starch degradation products, the so-called glucose syrups. These were originally based on partial acid hydrolysis of starch, usually corn starch. The modern technology of the production of these syrups has become increasingly sophisticated and both acid and enzymatic treatments are now involved in their production.

The products are usually classified according to the 'dextrose equivalent', where the reducing sugars are measured and expressed as dextrose. This value is calculated as a percentage of the dry substance. Thus a syrup with a high dextrose equivalent has been subjected to a greater degree of hydrolysis than one of a lower dextrose equivalent.

These syrups usually contain glucose, maltose and a range of the higher maltose oligosaccharides, maltotriose, maltotetrose, etc. Some preparations may give a faint colour reaction with iodine in potassium iodide, although these are more usually considered as maltodextrins.

The solubility of these mixtures in aqueous ethanol at the concentrations used to extract sugars is somewhat variable and the material insoluble in alcohol is usually sticky and difficult to filter.

Analysis of Degraded Starches

The analysis of these products usually involves the measurement of dextrose equivalents, the measurement of total sugars and may involve the separation and measurement of the various carbohydrate species present.

Dextrose equivalents are usually measured by a modification of the Lane–Eynon reducing sugar procedure (Chapter 8 (Reducing sugar methods)). The reducing sugar as glucose is then expressed as a percentage of the solids, usually determined from the specific gravity of the syrup.

Free glucose is most conveniently measured by a specific enzymatic method and the AOAC suggest the glucose oxidase procedure (Chapter 8 (Biochemical procedures)).

Total glucose can be measured after dilute acid hydrolysis or enzymatic hydrolysis using an amyloglucosidase preparation. Some preparations, however, contain retrogradation products of glucose that are not hydrolysed by amyloglucosidase.

If the preparation does not contain interfering substances that would give extraneous colours with strong acid, a condensation reaction method of the anthrone type (Chapter 8 (Condensation reactions)) can provide a convenient method for measuring total glucose.

Separation of the various carbohydrate species can be achieved by chromatographic methods and paper chromatography can often provide a reasonably quantitative estimate of the distribution of the various species. The reducing value of the higher maltose homologues seems to be the same on a molar basis as maltose in the ferricyanide method, and this method can be used to analyse column effluents or paper eluates [95].

Chromatographic separation on charcoal columns has been achieved and extensions of the ion-exchange procedures for sugar mixtures described in Chapter 8 (Ion-exchange separation of sugar mixtures) also provide a method for the quantitative analysis of mixtures of this kind.

These techniques apply in the main to the preparations themselves, the glucose or corn syrups. The analysis of these mixtures when part of a food containing other carbohydrates usually requires some extension of the procedures described in the earlier chapter on the analysis of sugar mixtures.

Some alternative approaches are given in the final section of Chapter 3; the prime requirement here is detection of these materials in a foodstuff and this will usually follow from the qualitative examination of the aqueous ethanolic and aqueous extracts. These will show whether or not higher maltose oligosaccharides are present. Comparison of the ethanolic and aqueous extracts can show whether alcohol-insoluble components are present. Dilute acid hydrolysis (0·4N H_2SO_4 for 1 h at 100 °C) is usually sufficient to convert the oligosaccharides to glucose, and from thereon a range of suitable methods is available. This technique may not be acceptable in the presence of lactose, which would also be hydrolysed or sucrose, where the fructose moiety may be degraded. In these circumstances enzymatic hydrolysis with an amyloglucosidase preparation may provide the most specific analytical method.

MODIFIED STARCHES

A wide range of modifications of starches has been developed to meet the special technological needs of the food processor [56]. The modifications are aimed at producing alterations in the starch molecule so that its behaviour under specific conditions is predictable and controllable. The properties most desired concern the resistance of the starch gel to retrogradation and its ability to form gels at low temperatures.

Heat treatment produces 'pregelatinised' starches that are soluble in cold water. These starches are used in the production of instant desserts, where the food is reconstituted with cold water or milk, and by manufacturers who wish to thicken a product that is mixed without heating.

Chemical treatments range from a very mild acid hydrolysis, which tends to attack the amylopectin molecule at the branch points, leading to a starch with a higher proportion of linear molecules, to extensive chemical modification of the starch molecule. These 'acid-thinned' starches swell less than the parent starch on cooking and produce a more fluid starch paste. Their major use is in the field of sugar confectionery, often in conjunction with other colloids in the production of gums, pastilles and jellies.

Further hydrolysis produces the maltodextrins with a dextrose equivalent of up to 20. These are used where soluble, non-sweet carbohydrates are needed.

Oxidation with hypochlorite, or more rarely permanganate or periodic acid, produces another range of starches. The reaction seems to involve oxidation of the primary hydroxyl group to carboxyl. This inhibits the tendency of the starch molecules to aggregate and leads to the production of clear gels. These oxidised starches are used in sugar confectionery and in food processing where clear gels are required. The gels formed by these starches tend to be soft.

The introduction of ester and ether groups is one of the more important areas of starch modification. Two broad types of product are produced; 'monostarch' derivatives, where the reagents react with the individual molecule to form 'side chains' and 'distarch', where a bifunctional group is used to produce cross-links or bridges between different molecules.

Distarch Derivatives

These are the more important as food additives. The cross-linking produces a starch that forms a paste which is stable at high temperatures and in acid conditions. These products are therefore suitable for use in fruit-pie fillings and soup-making and other high temperature processes.

Phosphate ester cross-links are introduced by reaction with phosphorous oxychloride and sodium metaphosphate in alkaline conditions. Adipic acid is also used to prepare ester distarches.

Monostarch Derivatives

These are most often used in association with distarch derivatives. The introduction of the ester or ether 'side-chains' inhibits the tendency of the starch chains to aggregate and gives the gel clarity and strength.

The types of ester used most commonly are the acetates and phosphates; ethylene oxide and propylene oxide are used in alkaline conditions to produce the corresponding ethyl and hydroxy-propyl ethers.

Analysis of Modified Starches

Many of the starch derivatives behave as natural starch in standard analytical procedures and thus present few problems to the food analyst.

Identification of the pregelatinised and partially hydrolysed starches depends on comparison with the composition of corresponding native starches.

Identification of the chemically modified starches depends on the detection of the organic groups that have been introduced into the starch molecule. Gas–liquid chromatography of the products of hydrolysis provides a way of doing this, particularly if it is coupled with a mass spectrograph.

These modifications affect the solubility of the starch but acid hydrolysis and reducing sugar methods will normally not distinguish between the modified starches and the other starch in foods. The introduction of cross-links into the starch affects the extent of enzymatic hydrolysis, for example the susceptibility of some modified starches to amyloglucosidase hydrolysis is much reduced.

The starch in foods containing these modified starches may therefore be underestimated if enzymatic hydrolysis or enzymatic methods for the glucose formed on hydrolysis are used.

CONCLUSIONS

Choice of Method in the Analysis of Starch

The fundamental question that determines the choice of the appropriate method for starch in foods concerns the definition of starch.

Polarimetric methods depend on the assignment of a specific rotation for starch and, while these methods are suitable for the analysis of one type of food where the characteristics of the starch are well-defined, their general use for unknown or less well-characterised starches is likely to be prone to error. If used for foods containing a mixture of starches or in which the starch has undergone some modification either involving heating or of a more extensive kind, the methods are likely to be unreliable.

Total acid hydrolysis will measure all the acid-hydrolysable polysaccharides in the food, including those derived from the non-cellulosic polysaccharides. The use of a glucose-specific enzymatic method for the analysis of the hydrolysates increases specificity but will also measure β-linked glucans.

Enzymatic methods are the most specific for measuring 1,4 and 1,6 α-glucans, which appears to be the most reasonable definition of starch. For nutritional purposes the choice of method must be the one that most closely simulates the extent of hydrolysis and utilisation in the animal (or man). There is quite good evidence that enzymatic methods meet this criterion most closely [21], although detailed experimental proof of which is the best enzyme preparation to use is not yet available.

The analyst must at this stage recognise that values for starch, like so many of the other food carbohydrates, are largely defined by the method of analysis used.

CHAPTER 5

The Measurement of Unavailable Carbohydrates: Structural Polysaccharides

The earlier chapters have been concerned mainly with the carbohydrates in foods which fall within the *available* category. This and the following chapter are concerned with the *unavailable carbohydrates*. This category comprises the polysaccharides that are not hydrolysed by the endogenous secretions of the mammalian digestive tract. Considerable degradation of many of these polysaccharides takes place in the rumen; the large intestine and the caecum of the non-ruminant herbivore and in the large intestine of man and other omnivores, as a consequence of the activities of the microflora of the contents of these organs. The products of this microbial activity are short-chain fatty acids. The term *unavailable carbohydrates* signifies that they are not sources of carbohydrate to the host animal although they are, of course, the major energy-yielding components in the diet of many animals.

The unavailable carbohydrates in foods may be considered as two major

TABLE 5.1

A CLASSIFICATION OF THE UNAVAILABLE CARBOHYDRATES IN FOODS

Principal sources in the diet	Description	Classical nomenclature
Structural materials of the plant cell wall	Structural polysaccharides	Pectic substances Hemicelluloses Cellulose
	Non-carbohydrate constituents	Lignin Minor constituents
Non-structural materials, either found naturally or used as food additives	Polysaccharides from a variety of sources	Pectic substances Gums Mucilages Algal polysaccharides Chemically modified polysaccharides

61

groups, depending on their sources; first those derived from the structural components of the plant cell wall and second the non-structural polysaccharides, gums, mucilages, algal polysaccharides and modified polysaccharides. This grouping is a little arbitrary and there are structural similarities in the polysaccharides in the two groups. The classification is summarised in Table 5.1.

The table lists the polysaccharides according to their classical names and includes the non-carbohydrate components, *lignin*, and the other minor constituents of the plant cell wall. While lignin is not a carbohydrate, it is important to consider its properties in relation to the measurement of cellulose.

This chapter is concerned with the measurement of the polysaccharides derived from the plant cell wall. The non-structural polysaccharides are considered in Chapter 6.

PHYSICAL STRUCTURE OF THE PLANT CELL WALL

The typical plant cell wall consists of a number of more or less distinct layers [82]. The first visible structure that appears in the development of the wall is the *cell-plate* which, on histo-chemical evidence, is rich in uronic acid polymers; this extends to meet the pre-existing walls and becomes the middle lamella in the fully formed wall.

The primary cell wall forms on this middle lamella; in this wall the cellulose fibrils are laid down in an apparently random network surrounded by a matrix of hemicellulose-type polysaccharides. The primary walls of most food plants are delicate structures [127]. The secondary wall is deposited on the primary wall in a number of distinct layers in which the cellulose fibrils are deposited in a parallel fashion, again in a matrix of hemicellulose. The secondary walls are usually much thicker than the primary wall and the rigidity of many plant structures is due both to the thick secondary walls of the collenchyma and to the turgor of these cells.

The process of lignification, in which lignin is deposited in the matrix of the wall, starts in the middle lamella, and usually at the junction of several cells. The process continues towards the interior of the cell, and the cells with completely lignified walls are usually dead. Lignification is an infiltration and the wall tends to swell during the process [31].

The distribution of lignin is very localised in the plant, being confined mainly to the xylem elements of the vascular bundles. In most leafy and root vegetables, and in the majority of fruits, the amount of lignified tissue is quite small. Seed coats and other specialised structures can be heavily lignified and the stones of many fruits are especially heavily lignified.

Some fruits contain lignified structures in their flesh, the pear for example, where small clusters of lignified stone cells are responsible for the gritty texture of pear flesh [31].

Grasses in general tend to rely on lignified supporting tissues to maintain the rigidity of their leaves and stems. The walls of these plants also contain appreciable amounts of silica.

The external surfaces of many plants are cutinised. The presence of cutin in these walls may make the cellulose less accessible to reagents, and this insolubility of cutin may lead to elevated values for lignin [121].

CHEMISTRY OF THE POLYSACCHARIDE
COMPONENTS OF THE PLANT CELL WALL

Nomenclature

The classical nomenclature for the components of the plant cell wall is derived from the methods used in the fractionation of the plant cell wall. The definition of these terms is probably more dependent on the method of isolation than on chemical structure.

Detailed studies of the chemical structure of the fractions isolated by classical means have shown that a range of structures are to be found in all the fractions, and these will be discussed later.

At this point it is important to emphasise that most of the knowledge relating to the structures of the polysaccharides in the plant cell wall is derived from detailed studies of the cell walls of woody tissues.

Detailed studies of the cell walls of what might be regarded as foods have been much more limited, although recently the interest in these substances has been increasing. However, the studies that are available suggest that deductions from observations on woody tissues have some general value.

The relationship between the classical nomenclature and the procedures used to isolate the components of the plant cell wall are illustrated in Table 5.2.

Pectic Substances

This group includes the polysaccharides that can be extracted with hot water [6]. It is usual to add a chelating agent such as EDTA or ammonium oxalate to the extraction medium in order to release those pectic substances present as the calcium salts.

The preparation of the other components of the plant cell wall is usually preceded by delignification, and this is most commonly achieved by treating the sample with chlorine: this results in the oxidation of the end-groups of many polysaccharides and some polysaccharides dissolve with the lignin

TABLE 5.2

TERMINOLOGY AND THE PROCEDURES USED TO ISOLATE THE COMPONENTS OF THE PLANT CELL WALL

Components	*Summary of extraction procedure*
Pectic substances	Extraction with hot water with the addition of a chelating agent such as ammonium oxalate or EDTA
Hemicelluloses (A and B)	Extraction with dilute alkali under nitrogen (A) Precipitated on neutralisation (B) Precipitated on neutralisation and addition of alcohol
Hemicellulose (C)	Extraction with strong alkali under nitrogen
Cellulose	Residue insoluble in 17·5% NaOH (α-cellulose)
Lignin	Residue insoluble in 72% (w/w) H_2SO_4 (Klason Lignin)

[23]. Some authors feel that the polysaccharides dissolving at this stage have the same composition as those left insoluble, but this is difficult to prove and seems intrinsically unlikely.

Hemicelluloses [113, 114]
These are extracted after removal of the pectic substances by extraction with dilute alkali. In many schemes of fractionation these are separated into two classes; hemicellulose A, which precipitates on neutralisation of the extract, and hemicellulose B, which precipitates on further addition of alcohol (to around 80%, v/v) [17, 18, 87].

Cellulose
This is the residue insoluble in strong alkali (17·5% (w/v) NaOH), which is usually called α-cellulose. Glucose polymers of a lower molecular weight than celluloses are found in the alkali extract, and are presumably formed by degradation of the native cellulose.

Structure

The classical fractions of the plant cell wall contain a range of polysaccharide structures. Moreover, there is considerable evidence that the fractionation procedures result in extensive structural degradation of the native polysaccharides present in the plant tissues. This appears to be true for the pectic substances, where the presence of labile arabinofuranosyl linkages means that extensive fragmentation occurs even under mild conditions. Aspinall [6] has suggested that the various types of polysaccharides seen in pectic fractions may be derived from a hypothetical native pectic polysaccharide. The alkaline extraction of the hemicelluloses also produces progressive degradation, and it has been suggested that it is more correct to regard the polysaccharides in the cell wall as falling into two categories, cellulosic and non-cellulosic; the latter category comprising a spectrum of polysaccharide structures [2]. This spectrum ranges from those rich in uronic acid residues (the pectic substances) to those low in uronic acid residues (the hemicelluloses). This method of classification is the most useful for the food analyst at the present time.

The types of structure found in the polysaccharides from the plant cell wall are summarised in Table 5.3. The table gives the main structural types,

TABLE 5.3

SUMMARY OF STRUCTURAL FEATURES OF THE UNAVAILABLE CARBOHYDRATES FROM THE PLANT CELL WALL

Primary grouping	*Main structural types*		*Structural variations*
Cellulose	Long chain 1,4-β-D-glucans		Degree of polymerisation
			Distribution of non-crystalline regions in the fibrils
			Presence and distribution of other sugars
Non-cellulosic polysaccharides	Galacturonans		Degree of polymerisation
	Galacturono-rhamnans		Extent of methyl esterification of uronic acids
	Arabinans		
	Xylans	—glucurono	Presence of acetyl groups
		—arabino	Number and distribution of side-chains
	Mannans	—galacto	
		—gluco	Type and extent of branching
	Galactans	—arabino	Intermolecular binding
	Glucans	β 1,3, 1,4	
		—xylo	

together with the main variants that have been observed in polysaccharides isolated from a range of plants.

Cellulose

This is the major structural polysaccharide in the plant cell wall [97]. In most plants a single type of polymer is present, a 1,4-β-D-glucan with a high degree of polymerisation. Cellulose is present in the cell wall in the form of fibrils, and these fibrils have a highly ordered structure with long regions possessing crystalline characteristics. The configuration of the cellulose chain permits a very close packing of the chains within the fibril and there is a high degree of intermolecular hydrogen bonding. Disordered, non-crystalline regions occur at irregular intervals in the fibrils.

Traces of other sugars are usually found in hydrolysates of most cellulose preparations: it is possible that these sugars occur in the non-crystalline regions and are, in part, responsible for the disordering.

The susceptibility of a cellulose preparation to either acid or enzymatic hydrolysis seems to be related to the extent of the non-crystalline regions, probably due to the fact that these regions, because of the reduction in hydrogen bonding, are more accessible to the hydrolytic reagents. Cellulose preparations can be extensively degraded by chemical treatments such as hot alkaline extraction in the presence of oxygen and, to some extent, by rigorous physical procedures such as ball-milling. Any procedure that either reduces the degree of polymerisation or dissociates the individual cellulose molecules and reduces intermolecular bonding will therefore alter the properties of the intact cellulose with regard to hydrolysis, and the analyst must take recognition of this fact in the choice of methods.

Non-cellulosic Polysaccharides

Much of the detailed structural work on these polysaccharides has been done with preparations obtained by the classical fractionation schemes. One is therefore compelled to use the terminology of these schemes in any discussion of their structures.

Pectic Substances

These water-soluble polysaccharides are characterised by the presence of a high proportion of galacturonans or galacturonorhamnans, although arabinans, galactans and arabinogalactans have been isolated from the pectic fraction of many plants. The characteristic pectic substance is often thought to be a homopolymeric galacturonan of the type found in sunflower seed heads. However, this view is not supported by the majority of other studies, which suggest that in the plant itself there is a complex heteropolymeric substance that is degraded during isolation [6].

The galacturonan chains contain 1,4-β-galacturonyl residues and, in most cases, rhamnose residues occur at intervals. The carboxyl groups of the uronic acids are esterified with methoxyl groups. The extent of this methyl esterification has profound effects on the physical properties of the molecule, in particular its gelling properties. Highly methylated pectic substances are termed *pectins*, whereas *pectinic acids* are only partly methylated and the term *pectic acids* refers to preparations that are devoid of methyl groups. The methoxyl groups can be removed by acid, alkaline or enzymatic hydrolysis either completely or partially, although this hydrolysis is usually accompanied by some degree of depolymerisation [9].

The other polymers found in the pectic fraction are, L-arabinans, which appear to be linear 1,5α chains with 1,3 single arabinose side chains; D-galactans as linear 1,4,β galactopyranosyl chains, and L-arabinogalactans with linear 1,4,-β-D galactopyranosyl chains with arabinose side chains.

Hemicelluloses

The material extracted by dilute alkali from most plant tissues is a complex mixture of polysaccharides, most of which are complex heteropolysaccharides containing between two and four different sugar residues. The fraction precipitating on neutralisation of the extract, *hemicellulose-A*, usually contains xylans with some uronic acid residues, whereas *hemicellulose B* preparations are more complex and contain both linear and highly branched polysaccharides [46].

Most of the structural information on this group of polysaccharides is derived from studies with woody tissues. An excellent and comprehensive review of this field has been published by Timell [114, 115]. The major types of polysaccharide present are *xylans*. These appear to be based on 1,4-β-D-xylan chains carrying side chains of several different kinds, 4-O-methyl-α-D-glucurono- and L-arabino- being the most common. Many cereals contain arabinoxylans, and they are therefore a common component of the non-cellulosic polysaccharides in foods [27, 85]. The arabinose side chains appear to be distributed at random along the xylan chains and are usually only a few residues long. The xylan chain itself is often branched.

Mannans frequently occur in the hemicellulose fraction; the homopolymeric substance is probably rare and seems to be a storage polysaccharide. *Gluco-* and *galacto*mannans are quite common, especially in preparations from seeds where they too may be storage forms [5]. The glucomannans appear to be linear, whereas the galacto- residues are usually found as side chains, often on glucomannans. The mannan chain has some of the conformational properties of the glucan chain in cellulose, and unsubstituted mannans and glucomannans are only soluble in strong alkali. The presence of galacto- side chains increases the solubility of the molecule, presumably because it interferes with intermolecular bonding. Substituted galactans of the *arabinogalactan* type are present in many hemicellulose preparations. These appear to be highly branched compounds with a 1,3-β-D-galactopyranosyl chain carrying galactosyl and arabinosyl side chains.

Many cereals also contain *β-glucans* with both 1,3 and 1,4 linkages. These substances are water-soluble and form part of the so-called 'cereal gums'; as such they are not hemicelluloses in the strict sense but it is convenient to consider them under this heading.

Lignin

Although not a carbohydrate, it is necessary to comment briefly on the structure of lignin at this point because of the implications of the presence of lignin in a food on the analysis of the polysaccharides. There appears to be a range of structures present in the lignin fraction and probably a number of different lignins occur in nature. The molecule is built up by the condensation of phenolic alcohols and is essentially an aromatic polymer containing a number of functional organic groups. It is extremely resistant to both chemical and enzymatic degradation and is possibly the most resistant substance found in nature. It is slowly soluble in organic solvents such as dioxane but is classically isolated as the residue insoluble in 72 % (w/w) H_2SO_4. Lignin fractions prepared in this way contain cutin and suberin and for this reason many values for lignin in foodstuffs are too high. In addition, protein–carbohydrate interactions occur when foodstuffs are heated to give materials that also analyse as lignin [120].

Lignin is one of the most difficult components of the plant cell wall to measure adequately but at the present time the procedure suggested by Van Soest and Wine [122] seems to be the most useful.

ANALYTICAL APPROACHES TO THE ANALYSIS OF THE STRUCTURAL CARBOHYDRATES

The analysis of the structural carbohydrates in foods presents considerable problems to the intending analyst. The complexity of the mixture and the range of structures involved implies that a detailed comprehensive analysis would be time-consuming and involve a considerable amount of fractionation.

Two types of approach have been used. The first, developed by those concerned with animal nutrition, can be called the *fibre school* and the second, arising from studies of the plant cell wall, can be called the *cell wall school*.

In the past the food analyst has been almost completely involved with the former and this approach must be considered first.

The Measurement of Fibre

Methods for the measurement of the indigestible residue of animal feeds were developed in Germany in the second half of the nineteenth century [54]. The procedure called the Weende method (after the place where it was developed) was based on treatment of the food with hot acid and alkali. The residue obtained by this method came to be called *fibre* or *crude fibre*. From its introduction this method was empirical and did not purport to measure any specific class of chemical substance, merely the indigestible residue in a food. However, the method became widely accepted, and the principles employed are still the basis of the method used at present.

Because of its empirical nature, close attention to experimental conditions is essential and numerous attempts have been made to improve the reliability of the method [52].

As a method for food carbohydrate it is possibly out of place here, as it does not measure any specific carbohydrate or group of carbohydrates; also it has been shown on several occasions that considerable losses of cellulose can occur [14]. Therefore, it does not measure the cellulose and lignin in a food with any accuracy, although this is often inferred.

The alkaline stage of the crude fibre procedure was believed to be the major source of variability and lack of precision in the method, and a modified procedure using acid digestion alone was suggested [36, 53, 126]. However, the precision of this method was dependent on the use of a detergent in the digesting acid. The work of Van Soest and his colleagues [119–122] in developing the use of detergents in the analysis of fibre provided the greatest advance in this field, and provided methods that could be used for the measurement of total cell wall material and fibre (the cellulose + lignin). These techniques, although still empirical, provide the best methods available for measuring the fibre in a food, especially plant foods such as animal forages [49].

In principle, the total cell wall material is measured as the residue after extraction with neutral detergent solution. The neutral detergent fibre

(NDF) contains all the cell wall except the water-soluble components, and includes a trace of protein associated with the structure of the wall. The acid detergent fibre (ADF) method involves extraction with boiling N sulphuric acid containing cetyl trimethyl ammonium bromide; the residue, acid detergent fibre, contains the cellulose and lignin in the food, together with some inorganic material, and can be used for the measurement of these constituents.

An estimate of the hemicelluloses in a food can be obtained by deducting the ADF value from the NDF value.

These procedures were developed originally for the analysis of forages, and their direct application to foods rich in fat, protein and starch can involve problems. Some starch is insoluble in the neutral detergent and will lead to elevated values unless it is first removed enzymatically [112]. The protein and fat seem to produce technical problems with regard to foaming during extraction and at the filtration stage. In general we have found it better to extract the lipid before attempting to measure either NDF or ADF in mixtures rich in fat.

The details of the crude fibre and the two detergent fibre procedures are given in Chapter 8 (Fibre methods). The former is included only for the sake of completeness, because until the ADF method becomes accepted as the standard procedure for fibre the analyst may still be called on to use the older and much less satisfactory method.

Measurement of Cell Wall Polysaccharides
The majority of work in this field has been undertaken as part of research work on the biochemistry of the plant cell wall or on the nutritive value of the components of the wall. For this reason most of the analytical methods used are time-consuming and not readily applicable to routine analytical work.

A large number of analytical schemes have been used [12, 45, 96, 123] and most workers have found it necessary to develop specific procedures for each type of tissue. Siegel [99] feels that a unified scheme of analysis that is applicable to all types of tissue may represent an attainable goal.

The main stages of the analytical schemes used in this type of work are summarised in Table 5.4.

Preparation of Sample
The sample must be air-dry and dehydration by freeze-drying or with acetone is preferable to oven-drying, as this may lead to the production of condensation artefacts. The degree of subdivision of the sample is

TABLE 5.4
STAGES IN THE ANALYTICAL FRACTIONATION OF THE PLANT CELL WALL

(1) Preparation of sample
(2) Extraction of sugars, lipids and pigments
(3) Extraction of water-soluble polysaccharides
(4) Delignification
(5) Extraction and fractionation of alkali-soluble polysaccharides
(6) Extraction of cellulose

important and some grinding is usually required but care is necessary to avoid excessive heating during this process. Ball-milling will produce a very fine powder from most tissues but should be used with moderation, as cellulose can be degraded by excessive milling.

Extraction
It is usual to extract the sample to remove free sugars, lipids and pigments, which may interfere with the subsequent extraction. Extraction with aqueous alcohols is commonly used and the residue obtained by a procedure of the type described earlier (Chapter 3) is very convenient to use for the analysis of the unavailable carbohydrates in general. A lipid solvent such as diethyl ether can also be used after the alcohol, as this will remove any residual lipid and some pigments; it is also readily removable from the residue. When analysing an unfamiliar material it is essential to check that polysaccharides have not been extracted into the aqueous alcohol; adjustment of the extract to 90% (v/v) ethanol will precipitate virtually all polysaccharides.

Extraction of Water-soluble Polysaccharides
This is usually carried out with hot water (90 ~ 100°C) or hot aqueous ammonium oxalate (0·5%, w/v) or a solution of EDTA. After cooling the extract is slightly acidified and the pectic substances are then precipitated by the addition of ethanol or acetone.

Delignification
The polysaccharides in the cell walls of woody tissues cannot be extracted completely without delignification and this also appears to be true for many grasses. The usual procedure for removal of lignin is treatment with chlorine or sulphite. A convenient delignification procedure for plant materials has been described by Whistler et al. [131]. Delignification is a very vigorous chemical procedure [28] and some modification of the insoluble

polysaccharides undoubtedly takes place, whilst some polysaccharides dissolve in the reagents [23]. It does not appear to be necessary for only lightly lignified tissues [129] and one would hesitate before carrying out the reaction on a sample rich in starch and protein (as one would have in the alcohol-insoluble residue from a human mixed diet). The product of delignification is referred to as *holocellulose.*

Extraction of the Hemicelluloses with Alkali

The extraction can be carried out with a range of strengths of alkali, although 5 % and 24 % (w/v) KOH and 4 % and 10 % (w/v) NaOH are quite common [45, 87]. It is essential to carry out the extraction in the absence of oxygen to avoid extensive degradation and it is usual to perform the extractions under nitrogen.

Prolonged and repeated extraction is necessary and with some materials intensely coloured extracts are produced.

The hemicellulose-A fraction is recovered by making the extracts slightly acid with acetic acid, and the hemicellulose-B fraction is precipitated by the addition of four volumes of ethanol.

The precipitated fractions can be separated further into linear and branched polysaccharides by formation of the iodine complexes with the linear polysaccharides [46]. Fractionation by precipitation with ethanol or by gel-filtration has also been used in a number of cases.

Extraction of Cellulose

The residue insoluble in strong alkali is often designated α-*cellulose* and β-1,4-glucan is the major component of the residue from many plant tissues. Traces of other sugars can usually be detected in the hydrolysates of α-celluloses and these may be derived either from contaminating hemicelluloses of the xylan type or from the cellulose molecules. Whatever their source, however, it is reasonable to suppose that they are derived from material that is intimately associated with the fibrils in the wall. The cellulose in this residue may be dissolved out with 72 % (w/w) H_2SO_4, and this provides a convenient way of measuring the polysaccharides in this residue.

Measurement of the Fractions

The precipitated fractions have been measured gravimetrically but this has some defects as an analytical method. In the first place, the precipitates usually form as fine gelatinous masses that are difficult, although not impossible, to recover quantitatively. A more serious objection, however, concerns their purity. The hemicellulose fractions are frequently associated with some protein and the α-cellulose fraction includes cutin, together with traces of lignin and inorganic matter. In most studies it is usual to hydrolyse the fractions and measure the component sugars. Many of the cell wall polysaccharides undergo complete hydrolysis in mineral acid, although some hydrolytic destruction occurs. No generally applicable conditions can be laid down for all polysaccharides and the analyst must establish the best conditions for each material. Hydrolysis in $0.4N$ or $N H_2SO_4$ seems to provide a reasonable starting point in these studies. Hydrolysis with trichloroacetic and trifluoroacetic acids [3] has also been used and the latter procedure provides a convenient starting point for the preparation of the alditol acetates for gas–liquid chromatography. The analysis of the sugars in these hydrolysates is most conveniently made by ion-exchange chromatography (Chapter 8 (Ion-exchange separation of sugar mixtures)) or gas–liquid chromatography (Chapter 8).

Uronic acids in these polysaccharide fractions can either be measured directly or after hydrolysis. Enzymatic procedures for the hydrolysis and analysis of pectic substances have been described [34] but hemicellulase preparations are not available at present and most cellulase preparations are of such low activity that their use in analysis is limited.

Place of 'Cell Wall' Methods in Food Analysis
The complexity of an analytical procedure for the structural polysaccharides in foods using the cell wall methods is immediately apparent and in general they can only provide the reference methods by which simplified procedures can be judged.

Many of the procedures have been developed for plant tissues and their direct application to many foods eaten by man is difficult.

CHOICE OF METHODS FOR THE ANALYSIS OF STRUCTURAL POLYSACCHARIDES

The ideal procedure for the analysis of the structural polysaccharides in foods would involve the separation and analysis of all the various classes of polysaccharide likely to be present in foods [106].

Such an ideal is impracticable at present, although the development of high-pressure chromatography may provide a means for the separation of the non-cellulosic polysaccharides.

Some simplification of the fractionation and analysis is required for the practical analyst. A simplified procedure for unavailable carbohydrates was developed during the course of studies on the 'digestibility' of the energy-yielding components of the human diet [104, 108]. The basis of this method was to use an alcohol-insoluble residue and to remove the starch by enzymatic hydrolysis. The non-cellulosic polysaccharides were separated into water-soluble and water-insoluble fractions and measured as their component sugars after dilute acid hydrolysis. Cellulose was measured by dissolving the cellulose in 72% (w/w) H_2SO_4 and measuring the glucose colorimetrically. This method has been used on a variety of foods and appears to give values that are related to values obtained by the application of the more detailed 'cell wall' methods. The values for cellulose and lignin obtained by this method also agree with those obtained from ADF by the Van Soest procedures [76, 93].

The methods described in the selected methods (Chapter 8) include a detailed method of the 'cell wall' type as developed for grasses and a simplified method for unavailable carbohydrates.

CHAPTER 6

The Measurement of Unavailable Carbohydrates: Non-Structural Polysaccharides

This chapter is concerned with the heterogeneous collection of polysaccharides found in many foods [48, 100]. The only feature they have in common is that they are not digested by mammalian digestive enzymes and they are therefore classified as 'unavailable'. Many of them are water-soluble and all are soluble in dilute alkali. Therefore, in most schemes for the analysis of the unavailable carbohydrates they will analyse as 'pectic substances' or 'hemicelluloses' or, according to the preferred terminology discussed in Chapter 5, as non-cellulosic polysaccharides. These non-structural polysaccharides frequently share important features of chemical structure with the non-cellulosic components of the plant cell wall [6].

The analysis of these polysaccharides depends on the qualitative identification of their presence in a foodstuff, followed in most cases by hydrolysis and measurement of the sugars thus produced. The specific analysis of these polysaccharides in the presence of other non-cellulosic polysaccharides is difficult and fortunately is rarely required.

The Official Methods of the AOAC [4] for these substances are restricted to the identification of the polysaccharides, and even these procedures are subject to qualification in the sense that a negative reaction is not regarded as conclusive proof that the polysaccharide in question is *not* present in the food. More specific methods would involve fractionation of the non-cellulosic polysaccharides or, in a few cases, hydrolysis and measurement of a specific unique constituent. No general methods are available and the chosen method of analysis must be based on the structure and properties of the individual polysaccharides.

This chapter will therefore concentrate on reviewing the types of polysaccharides of this category that are found in foods, and finally consider the most appropriate methods for the analysis of these components.

GUMS

The plant gums form a very heterogeneous group of complex, branched heteropolysaccharides. They are derived most commonly from exudates of plants, frequently at the site of some physical damage. They are in wide use in the food industry as additives but in some cases they are being replaced by modified starches [56].

In general the gums are water-soluble, although warm water may be required to extract them initially.

Structure

A wide range of structural types are seen among the plant gums and one of the most convenient methods of classification is based on the type of polysaccharide core chain present. It is possible, however, that the physical properties of the gum may be determined by the nature and configuration of the side chains.

The major structural types seen in the plant gums are summarised in Table 6.1. The most important gums with regard to food technology are gum arabic, gum tragacanth, khaya gum, and the seed gums from the locust bean and guar.

Gum Arabic

This has the longest history of use in food processing; in nature it is found as the neutral or slightly acidic salt of a complex polysaccharide. The main feature of its structure is an interior galactan chain, which is branched and carries arabinosyl side chains with some rhamnosyl groups. The side chains are terminated by glucuronic acid residues, which may be substituted at the 4-O position with methyl groups.

The gum is very soluble in water and solutions containing up to 50 % (w/v) can be prepared. The gum will thus produce a highly viscous gel on its own but is usually combined with other gums as a thickening agent.

It is insoluble in alcohol and partially soluble in aqueous alcohols up to ~ 60 % (v/v), when it is virtually insoluble. Solutions of the gum are usually slightly acidic and at this pH (4·5–5·5) undergo slow autohydrolysis of the arabinofuranosyl groups.

The presence of the gum in foodstuffs can be deduced by making a hot water extract of the food and then precipitating the polysaccharides from this extract with alcohol. The precipitate is completely soluble in water and gives a greenish brown colour with concentrated sulphuric acid.

Gum Tragacanth

This too has a long history of use in foods; it is prepared from the exudate of *Astragalus* spp. The detailed structure of the gum is still not clear but tragacanthic acid, a major component, has an interior core of a

TABLE 6.1
MAJOR STRUCTURAL FEATURES OF PLANT GUMS

Type	Core	Side-chains	Terminal groups	Examples
Galactan	D-galactopyranosyl 1,3 and 1,6	L-arabinofuranosyl some L-rhamnopyranosyl	D-glucuronic or 4-O-methyl glucuronic	Gum arabic[a]—and other *Acacia* spp. Asafoetida, Araucaria Lemon, Golden 'Apple
Glucuronomannan	4-O-substituted D-glucuronic acid alternating with 2-O-substituted D-mannosyl residues	L-arabinofuranosyl L-arabinopyranosyl D-glucopyranosyl D-galactosyl	Variable	Gum ghatti *Prunus* gums
Galactomannan	D-galacturonic acid L-rhamnosyl	D-xylopyranosyl L-fucopyranosyl D-galactopyranosyl arabinoxylo	Variable	Tragacanthic acid[a] Khaya gums[a] *Stercularia* gums
	D-mannopyranosyl D-galactopyranosyl	D-galactopyranosyl	—	Locust bean gum[a] Guar gum[a] (more side chains)
Xylan	D-xylopyranosyl	D-galactopyranosyl L-arabinofuranosyl	Variable	Sapote gum *Watsonia* gum
Xyloglucan	1,4 β-D-glucopyranosyl	D-xylopyranosyl	Variable	Tamarind

[a] Gums of particular importance in food processing.

galacturonan with a variety of side chains. The gum appears to contain a soluble portion that forms a colloidal solution and an insoluble portion that swells to give a gel. Highly viscous solutions containing about 1 % (w/v) can be prepared, whereas at 2 ~ 4 % (w/v) a thick paste results.

The solutions are usually acid (pH 5 ~ 6) and the viscosity of the solutions are reasonably stable in acid conditions.

The gum can be detected in the alcohol-precipitated material from the water extract by the blue colour formed with chlorozinc iodide and by the yellow colour formed by treatment with 10 % NaOH on a steam bath.

Gums Karaya and Stercularia
These are also of galacturonorhamnan type. Their structures are not completely established. The gums typically contain acetylated residues.

These gums absorb water to give viscous colloidal solutions. Heating produces clear colloidal dispersions and concentrations of up to 25 % (w/v) are possible; normally, however, a 3–4 % concentration is the maximum obtainable by cold water hydration. These gums have largely been replaced as food additives in favour of better stabilisers.

Locust Bean Gum (Carob Bean Gum)
This is derived from the seeds of *Ceratoria* sp. The gum is produced from the milled endosperm of the seed, which is then dehusked. Structurally this gum is a neutral galactomannan with every fourth or fifth manopyranosyl carrying a galactose side chain.

The gum is insoluble in cold water but dissolves on heating to give a very viscous solution. The solutions are neutral and the gum is stable over a wide range of pH values. It is most commonly used in conjunction with another additive such as carrageenan.

Guar Gum
This is also a galactomannan isolated from the seeds of the guar plant *Cyamopsis*. Structurally it differs from locust bean gum in that the galactose side chains are carried on alternate mannose residues in the core chain.

The gum hydrates rapidly with cold water to form a very viscous solution. The polymer is neutral and is stable over a wide pH range.

These galactomannan gums when precipitated with alcohol give a pink or red-brown colour with concentrated sulphuric acid.

Qualitative Identification of Gums
The AOAC propose a scheme for the identification of gums in foods based on a series of tests with iodine potassium iodide in zinc chloride, iodine in

TABLE 6.2

(1) Test precipitate with iodine potassium iodide in zinc chloride
    (a) Blue   (i) Warm with 10% NaOH on steam bath
                                           yellow colour ≡ Tragacanth
            (ii) Add iodine to 0·1N   *blue colour* ≡ Starch
            (iii) Test (ii) negative   ≡ Quince

(2) Test precipitate with alcoholic iodine
    (a) Opaque blue-black   (i) Stains with
                            Ruthenium red,   ≡ Agar
                            does not dissolve
                            or lose shape
                            with water
    (b) Brown or lilac   (ii) Stains blue with
                            alcoholic
                            methylene blue   ≡ Carrageenan

(3) Test precipitate with Ruthenium red
    (a) Pink granular mass   (i) Heat with HCl,
                            pink   ≡ Karaya ≡ Khaya

(4) Concentrated H_2SO_4, warm gently
    (a) Pink or red-brown   ≡ Carob bean
    (b) Greenish brown   ≡ Acacia

alcohol, Ruthenium red and concentrated sulphuric acid. These tests, taken in conjunction with other tests and the nature of the original alcoholic precipitate, give more or less definitive evidence of the type of gum present. These tests are summarised in Table 6.2.

MUCILAGES

The polysaccharides included in this category are rather ill-defined; they are found associated with many seeds and are water-soluble with strong water-binding properties. It is usual to divide them into two classes, *neutral* and *acidic*.

The neutral mucilages include galactomannans, glucomannans, arabinoxylans and, in one instance, xyloarabinans. The structures and properties of these polysaccharides, where they have been studied, are similar to the related polysaccharides discussed in Chapter 5.

The acidic components are mainly galacturonorhamnans. The properties of the acidic mucilages are comparable to related polysaccharides in the pectic group of substances and the major differences seem to be due to the extent of branching and methoxyl substitution.

Analytically the mucilages are measured with the other water-soluble, non-cellulosic polysaccharides.

ALGAL POLYSACCHARIDES [89]

A number of polysaccharides extracted from algae (usually marine forms) are widely used in food processing [136]. The preparations most widely encountered in foods, are the *alginates*, either as their salts or esters, and two of the sulphated polysaccharides, *agar* and *carrageenan*.

Alginic Acid and Alginates
Alginic acid can be isolated from many species of brown algae. In the plant the polysaccharide is present as a mixture of salts of Na, K, Ca and Mg and, as such, is insoluble. Extraction involves treatment with acid and extraction with sodium carbonate solution, from which it can be isolated by precipitation with alcohol or as the free acid or calcium salt. The free acid is insoluble and contains D-mannuronic and L-guluronic acids linked possibly in a linear unbranched molecule. The two uronic acids appear to occur in blocks interspersed with mixed regions and it is possible to obtain predominantly mannuronan and guluronan fragments from hydrolysates.

Alginic acid is soluble in aqueous solutions of the alkali metal hydroxides and carbonates; these salts give highly viscous solutions. Polyvalent ions precipitate their salts as hard gels or films. Esters with propane 1-2-diol are stable in acid solutions and are widely used.

The alginates have characteristic infra-red spectra, and methods for their identification by comparison of the spectra of alginate films prepared from foods have been developed [19].

Agar
Agar can be extracted from several species of algae. The polysaccharide is insoluble in cold water but soluble in boiling water, giving a clear solution which, on cooling, sets to a firm rigid gel. A $1 \cdot 5 \%$ (w/v) solution, for example, gels on cooling to 32–39 °C to give a gel that does not melt below 85 °C.

Agar consists of two components, *agarose* and *agaropectin*. Agarose is composed of D-galactose and 3,6 anhydro-L-galactose in an alternating chain. Side chains of 6-O-methyl D-galactose are also present. Agaropectin is probably a mixture in itself, containing D-galactose, 3,6 anhydro-L-galactose, half-ester sulphate and D-glucuronic acid. The agarose component seems to have the best gelling characteristics.

Carrageenan (Irish Moss)

This is prepared from the water extract of a number of red algal species. A number of different forms of carrageenan have been recognised.

κ-carrageenan is precipitated by KCl and contains 3,6 anhydro-D-galactose in contrast to the L-form in agarose. The molecule contains alternating 1,3 D-galactose and 1,4 3,6-anhydro-D-galactose residues, these residues carrying sulphate groups.

λ-carrageenan is a related type of structure but is much more highly sulphated.

Carrageenan forms salts and the properties of these salts have important effects on the solubility of the polysaccharide and its behaviour as an additive. It is frequently used in conjunction with another component such as carob bean gum.

The polysaccharides are strongly charged, have high molecular weights, and are capable of reacting with both large and small molecules.

MODIFIED CELLULOSES

Some cellulose ethers are used as food additives and their use is common in many so-called 'slimming foods'. The introduction of the methyl group greatly modifies the property of the cellulose molecule, it becomes more soluble and more easily hydrolysed by acid. A range of ether derivatives are available and the degree of etherification is an important factor in the choice of additive.

The methyl celluloses possess the unusual property of becoming less soluble as the temperature of the solution is raised and this property can be used to confirm the presence of this type of polymer [84].

Analysis depends on extraction with cold water and recovery as a precipitate on heating. The polysaccharides can be hydrolysed by acid and measured by either a reduction method or a suitably calibrated condensation method. Some methyl-celluloses appear to be hydrolysed by mixed enzyme preparations of the Takadiastase type.

CONCENTRATIONS IN FOODS

These polysaccharide food additives are present in a range of manufactured foods and their overall use is increasing as more becomes known about the relation between physical behaviour and chemical structure. The amounts in foodstuffs are usually quite low as it is possible to achieve the desired physical effects at these levels. Some values for the typical levels of alginates are given by Wylie [136] and the alginate concentration in the final product is of the order of $0.1 \sim 0.2\%$ for most stabilising uses, although it may rise as high as 2.2% when gelling and thickening properties are required. In general the levels of carrageenan used are somewhat lower and of the galactomannan group higher, but the actual level used by different manufacturers is variable and depends on both the precise effects required and on the properties of the preparation used.

The analyst must therefore regard the polysaccharides discussed in this chapter as forming a minor proportion of the total carbohydrate in a foodstuff.

CHOICE OF ANALYTICAL METHOD

There are few standard procedures for these substances [84]. Isolation from the food sample, although in essence this should be reasonably straightforward, may be complicated by interactions between protein and polysaccharide. Hot water extraction alone may be insufficient if insoluble salts have formed during the stabilisation of the gel in the food, and it may be necessary to add a chelating agent to the extracting medium.

The polysaccharides can be recovered from the aqueous solution by the addition of ethanol; some will precipitate at around 60% (v/v) but adjustment to 80% (v/v) is possibly safer.

If one type of polysaccharide species is present, then a conventional method for polysaccharides can be employed. Thus dilute acid hydrolysis and the measurement of a component monosaccharide would suffice if one had a standard of the relevant polysaccharide with which to calibrate the method.

In most cases a condensation colour reaction can be applied directly to aqueous solutions of these polysaccharides. Alginates have been measured directly in extracts of textile fibres by reaction with ferric chloride and orcinol. However, this procedure is not directly applicable to foods [22] and the uronic acid method described on p. 111 would be preferable.

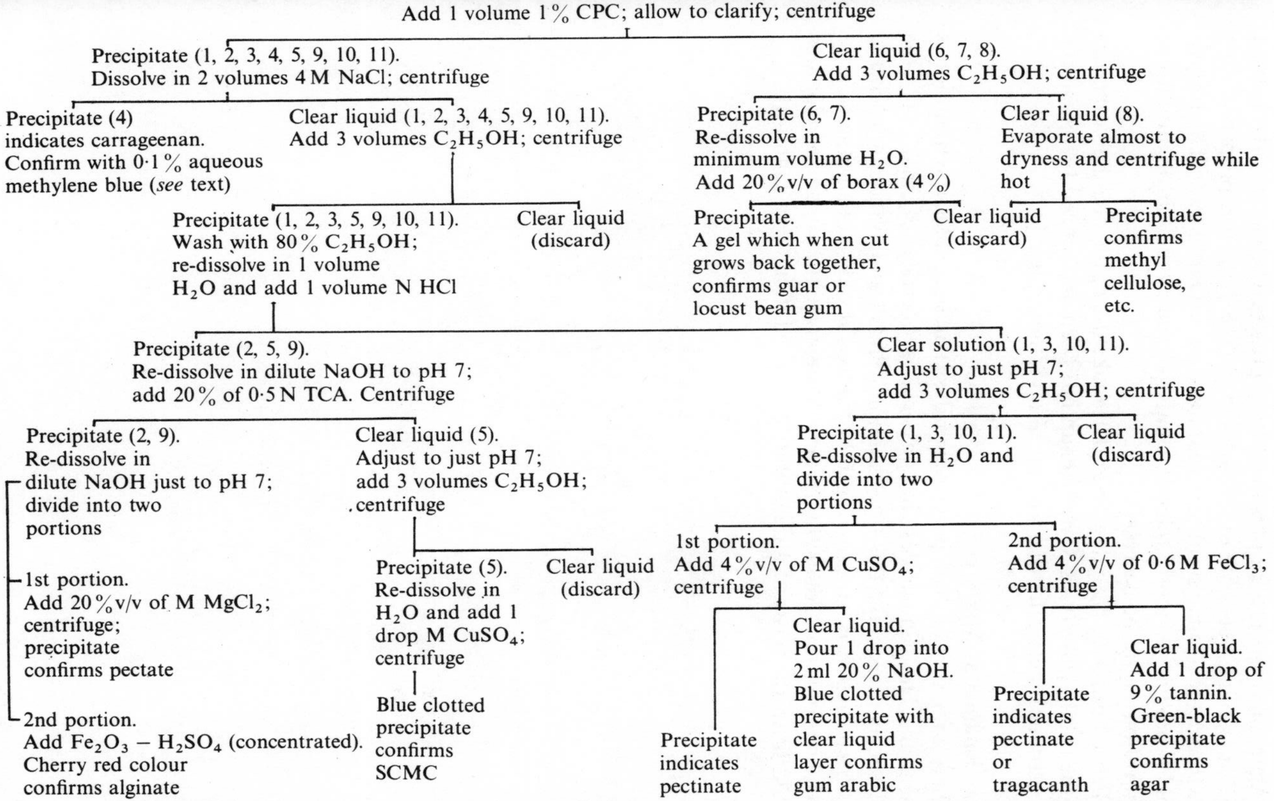

Add 1 volume 1% CPC; allow to clarify; centrifuge

Precipitate (1, 2, 3, 4, 5, 9, 10, 11). Dissolve in 2 volumes 4 M NaCl; centrifuge

Clear liquid (6, 7, 8). Add 3 volumes C_2H_5OH; centrifuge

Precipitate (1, 2, 3, 4, 5, 9, 10, 11) indicates carrageenan. Confirm with 0·1% aqueous methylene blue (*see text*)

Clear liquid (1, 2, 3, 4, 5, 9, 10, 11). Add 3 volumes C_2H_5OH; centrifuge

Precipitate (6, 7). Re-dissolve in minimum volume H_2O. Add 20% v/v of borax (4%)

Clear liquid (8). Evaporate almost to dryness and centrifuge while hot

Precipitate confirms methyl cellulose, etc.

Precipitate (1, 2, 3, 5, 9, 10, 11). Wash with 80% C_2H_5OH; re-dissolve in 1 volume H_2O and add 1 volume N HCl

Clear liquid (discard)

Precipitate. A gel which when cut grows back together, confirms guar or locust bean gum

Clear liquid (discard)

Precipitate (2, 5, 9). Re-dissolve in dilute NaOH to pH 7; add 20% of 0·5 N TCA. Centrifuge

Clear solution (1, 3, 10, 11). Adjust to just pH 7; add 3 volumes C_2H_5OH; centrifuge

Precipitate (2, 9). Re-dissolve in dilute NaOH just to pH 7; divide into two portions

Clear liquid (5). Adjust to just pH 7; add 3 volumes C_2H_5OH; centrifuge

Precipitate (1, 3, 10, 11). Re-dissolve in H_2O and divide into two portions

Clear liquid (discard)

1st portion. Add 20% v/v of M $MgCl_2$; centrifuge; precipitate confirms pectate

2nd portion. Add $Fe_2O_3 - H_2SO_4$ (concentrated). Cherry red colour confirms alginate

Precipitate (5). Re-dissolve in H_2O and add 1 drop M $CuSO_4$; centrifuge

Clear liquid (discard)

Blue clotted precipitate confirms SCMC

1st portion. Add 4% v/v of M $CuSO_4$; centrifuge

2nd portion. Add 4% v/v of 0·6 M $FeCl_3$; centrifuge

Precipitate indicates pectinate

Clear liquid. Pour 1 drop into 2 ml 20% NaOH. Blue clotted precipitate with clear liquid layer confirms gum arabic

Precipitate indicates pectinate or tragacanth

Clear liquid. Add 1 drop of 9% tannin. Green-black precipitate confirms agar

FIG. 6.1. Fractionation scheme for polysaccharide additives. (CPC = cetyl pyridinium chloride.) The numbers refer to the type of polysaccharide: 1, Agar; 2, Alginate; 3, Gum Arabic; 4, Carrageenan; 5, Na carboxymethyl cellulose; 6, Guar gum; 7, Locust bean gum; 8, Methyl cellulose; 9, Na polygalacturonate; 10, Pectin; 11, Gum Tragacanth. (Reproduced from *The Analyst*, **84**, with permission.)

Agar can frequently be presumed to be present when a hot water extract gels on cooling but some other polysaccharides have this property.

More detailed analysis of the sugar mixture produced by acid hydrolysis would provide a more reliable analysis and some chromatographic separation would be necessary.

Procedures have been described that make use of the binding properties of the sulphated groups. Carrageenan has been determined by measuring spectrofluorimetrically the binding of acridine orange [30].

Fractionation of Mixtures

Where a mixture of these polysaccharides is present, some fractionation is essential. The similarities in physical properties of many of the polysaccharides make this fractionation difficult and, although a scheme has been suggested recently (Fig. 6.1), it is so far only a qualitative scheme [84]. The application of methods for total sugars to the isolated precipitates could form the basis of a quantitative method if suitably calibrated.

CHAPTER 7

The Analysis of Carbohydrates in Specific Groups of Foods

In the earlier chapters the view was advanced that the analysis of food carbohydrates must be founded on a knowledge of the types of carbohydrate species present in the food being analysed.

This qualitative information can be established by the analyst if suitable qualitative procedures form part of the normal analytical scheme. In practice, however, the analyst will most commonly be concerned with a limited range of foodstuffs and, moreover, a range for which the qualitative composition is well-established.

The aim of this chapter is to draw together this qualitative knowledge and to consider this in the light of any special analytical problems posed by the foodstuff to an intending analyst.

MILK AND MILK PRODUCTS

These products contain a limited range of carbohydrates and in many of them lactose is the only carbohydrate of concern to the analyst.

The analysis of sweetened condensed milks involves the determination of lactose and sucrose. A few milk preparations contain starch hydrolysates of the maltodextrin category.

In the liquid products the carbohydrate is in solution and the only preparation of the sample required is the removal of interfering substances, a stage that usually involves deproteinisation.

Liquid Milk
Specific procedures for liquid milk are available where a range of deproteinising agents is used to provide a protein-free filtrate that is suitable for measurement of lactose by the standard reducing sugar methods (Chapter 8). These filtrates are also suitable for many of the condensation reactions.

Deproteinisation of milks from some species other than the cow may occasionally require some modification of the conditions due to variations in the proportions of different proteins present. The most frequently used methods of deproteinisation are those producing a bulky precipitate of an inorganic hydroxide, which causes the proteins to coprecipitate. However, perchloric acid deproteinisation is the method of choice where it is proposed to use an enzymatic method for the measurement of the lactose (Chapter 8 (Biochemical procedures)).

The concentration of lactose in the milks of different species is reasonably characteristic for that species and some typical values are given in Table 7.1. The lactose concentration seems to be less variable than the fat content of milks.

TABLE 7.1
CONCENTRATION OF LACTOSE IN MILKS OF DIFFERENT SPECIES (G/LITRE)

Cow	47	Camel	79
Goat	44	Rat	33
Ewe	44	Rabbit	19
Mare	60	Elephant	56
Sow	37	Hippopotamus	44
Human		70	

Sources [69, 109].

Small amounts of non-lactose carbohydrates have been found in milks; these appear to be amino sugars and substituted amino sugars. The amounts of them in most milks are insufficient to produce much error in normal food analysis.

Sweetened Condensed Milks

The analysis of these milks involves deproteinisation, usually after dilution, and then the analysis of a mixture of sucrose and lactose. Procedures involving the measurement of reducing sugars before and after inversion have been developed. The inversion can be carried out either with acid at room temperature or with invertase. Some correction for the effect of sucrose on the copper oxide reduced by the lactose is necessary and the reference tables (*see* Appendix) contain a column for lactose in the presence of sucrose.

Enzymatic analysis of lactose and sucrose mixtures can be carried out quite readily [8] and the two sugars can also be separated by ion-exchange chromatography, although this is usually not necessary as reliable values for sucrose/lactose mixtures can be obtained by the simpler procedures.

Dried Milks

The principles in the analysis of milk powders are similar to those for liquid milks. Lactose is poorly soluble in aqueous alcohols and it is usually better to dissolve the milk powder in hot water and then deproteinise the solution.

Milk powders containing lactose, lactose and sucrose, and lactose together with a range of other carbohydrates are found, and powders in the last category can present a complex mixture to the analyst. Separation of the mixture by ion-exchange chromatography or analysis by enzymatic methods is essential with the more complex mixtures. Analysis by gas–liquid chromatography of the trimethylsilyl derivatives of the mixture is possible but separation of the anomeric forms of lactose from sucrose is not complete and analysis before and after inversion with invertase is necessary.

Cheeses

Most cheeses contain only small amounts of sugars; extracts with water or aqueous alcohol usually contain small amounts of reducing substances but these are usually non-carbohydrate.

Whey cheeses of the *mysöst* type (brown soap cheeses) contain quite high concentrations of lactose, which can be measured after extraction and deproteinisation.

EGGS

Small amounts of reducing substances can be extracted from eggs and there are small amounts of glycoprotein present. For most analytical purposes, however, eggs can be considered to contain virtually no carbohydrate or possibly a trace, $\sim 0.1\%$.

MEAT, FISH AND THEIR PRODUCTS

Carcase Meats

These usually contain only traces of reducing substances, of which a small portion is carbohydrate.

Offals

The liver in the living animal may contain an appreciable amount of glycogen. On death this is rapidly degraded and, unless vigorous and rapid precautions are taken to inhibit enzyme activity, it is unusual to find much residual glycogen in a liver after removal from the animal and storage at room temperature. Significant amounts of carbohydrate can still be detected, however, some hours after death. It is rare for a food analyst to be asked to consider the glycogen in a meat but it can be measured as glucose after hydrolysis with amyloglucosidase or, if the need arises, after dilute acid hydrolysis.

Meat Products

Flour or other cereal fillers are widely used in the formulation of many meat products and these contribute minor amounts of sugars but significant amounts of starch. Free sugars can be estimated after extraction with aqueous alcohol but non-sugar reducing substances are often present and the more specific sugar methods are preferable. The starch in these products can be measured after hydrolysis of the alcohol-extracted material. Direct acid hydrolysis will usually give results of reasonable accuracy, provided the cereal filler present is a refined one. If a filler containing unavailable carbohydrates such as soya flour or an unrefined cereal is used, this procedure will overestimate the starch and an enzymatic hydrolysis is to be preferred.

The addition of milk powder to some meat products will contribute lactose to these foods and enzymatic procedures have been described for measuring lactose in mixtures of this type [7].

Fish

Fish flesh contains small concentrations of sugars, which are usually ignored in food analysis. Cereal fillers are used in the formulation of many fish products and the carbohydrates in these can be measured by the procedures outlined for similar meat products.

CEREALS

Grains and Flours [62]

Cereals contain small amounts of free sugars of the order of $1 \sim 2\%$, although this will rise if the grain has been allowed to germinate, for example in the preparation of malted cereals.

The free sugars are usually measured in aqueous ethanolic extracts. Glucose, fructose and sucrose can frequently be detected together with some maltose. Some care is necessary in the extraction if hydrolysis of the glucofructans found in most cereals is to be avoided [80].

The mixture of polysaccharides in cereals depends on the type of cereal and on its extraction rate. Cereal preparations of low extraction rate, where the branny outer layers and the germ have been largely removed, contain mainly starch, although small amounts of arabinoxylans are usually present.

The comprehensive analysis of cereal foods therefore requires the application of the complete range of procedures discussed earlier in general terms (Chapter 3 (Summary)). Several detailed schemes have been proposed and that of Fraser and his colleagues [43] is possibly the best tested, although this has its limitations where the unavailable carbohydrates are concerned.

The starch in low extraction rate flours can be measured by dilute acid hydrolysis with only minor inaccuracy.

Baked and Other Cooked Cereal Foods
These may contain added sucrose or glucose syrup in their formulation and the free sugars found in these products include glucose, fructose, sucrose and maltoses (if glucose syrups have been used). The amounts of free sugars present vary and some proprietary breakfast cereals and many sweet biscuits contain $20 \sim 25\%$ (w/w) free sugars. The starch in many products will have been subjected to fairly high temperatures during baking and appreciable degradation and dextrinisation is usual. Once again, these foods require the application of a complex analytical scheme for the comprehensive determination of all the carbohydrates present.

Extraction with aqueous ethanol provides a convenient way of isolating the free sugars; the residue insoluble in alcohol provides a convenient starting material for the estimation of the polysaccharides. Where the concentrations of unavailable carbohydrates are low, or a somewhat lower standard of accuracy can be tolerated, dilute acid hydrolysis followed by the measurement of glucose provides a reasonable estimate of the starch present. In foods where high extraction cereals or added fruit are present, enzymatic hydrolyses of the starch is necessary to avoid interference from non-cellulosic polysaccharides.

VEGETABLES

The plant organs that collectively form the vegetables are drawn from several distinct types of structure and it is more convenient to consider them separately [31], although in general the same analytical procedures can be used throughout.

Leafy Vegetables
These contain a low proportion of dry matter and the values for carbohydrate in the fresh materials reflect this fact. If results are expressed on a dry matter basis, these vegetables are largely carbohydrate, although protein is a significant component and may be present up to 20% on a dry matter basis.

The amounts of free sugars are small and typically glucose, fructose and sucrose are present [67, 102]. The proportions of the sugars vary from species to species: some typical values are given in Table 7.2. Starch can often be detected, although the amounts in general are small. The starch granules are localised around the stomata and distribution within a sample may be quite uneven.

The unavailable carbohydrates make up about $20 \sim 30\%$ of the total dry matter and all the structural carbohydrates of the plant cell wall are present. The pectic substances tend to be lower than in fruits and the amounts of lignin present are usually quite low.

Legumes
The more mature legumes contain a much higher proportion of solids, and in some of the dry seed legumes the solids may be comparable to that in cereals. The starch concentration in these mature products is also quite high, and may account for more than 50% of the dry matter. This high starch content has the effect of diluting the concentrations of sugars and unavailable carbohydrates present.

Root Vegetables
These have compositions very similar to those seen in the leafy vegetables, and the levels of free sugars and starch are usually quite low. Many root vegetables contain fructans, which in a few cases, the Jerusalem artichoke for example, account for most of the storage carbohydrate. These are readily hydrolysed and care is needed when the free sugars are extracted. Freezing and thawing can bring about the hydrolysis of these substances.

Stem Tubers: Potato
In the potato the water content is usually a little lower than in the true root vegetables and the concentration of starch much higher. The concentration of free sugars is low in the freshly harvested tuber and rises during storage [117].

TABLE 7.2

TYPICAL VALUES FOR THE CARBOHYDRATES IN VEGETABLES[a] (g MONOSACCHARIDES/100 g FRESH WEIGHT)

Vegetable	Total solids	Glucose	Fructose	Sucrose	Other sugars	Starch	Non-cellulosic polysaccharides	Cellulose	Lignin	Total dietary fibre
Leafy vegetables										
Broccoli	11~12	0.74	0.56	0.39	R[c]	0.2	2.7	0.83	0.03	3.56
Brussels sprouts	11~12	0.73	0.73	0.64	R	1.0	3.0	1.04	0.24	4.28
Cabbage	5~9	1.75	1.33	0.23	R	0.3	1.8	0.62	0.30	2.70
Cauliflower	7~9	0.92	0.81	0.56	R	0.4	1.0	1.07	Tr.	2.10
Lettuce	4~6	0.29	0.47	0.10	R	0	0.46	1.03	Tr.	1.50
Legumes										
Beans, haricot	89~91	—	—	—	—	46.2	19.8	4.8	0.8	25.4
runner	6~8	1.00	1.34	0.43	—	0.2	1.5	1.2	0.2	2.90
Peas	9~11	0.01	0.01	0.67	—	6.6	3.6	1.4	0.2	5.20
Root vegetables										
Carrot	11~12	1.04	0.92	0.89	—	0.1	2.2	1.48	Tr.	3.70
Turnip	6~7	1.50	1.18	0.42	—	Tr.[d]	1.5	0.70	Tr.	2.20
Fruiting vegetables										
Cucumber	3~5	0.86	0.87	0.06	—	Tr.	—	—	—	0.4
Pepper	6~7	1.08	1.01	0.13	—	Tr.	0.66	0.23	Tr.	0.9
Tomato	5~6	1.13	1.39	0.03	—	Tr.	0.61	0.50	0.29	1.3
Potato freshly harvested[b]	19~24	0.13	0.10	0.12	—	20.3	2.2	0.28	Tr.	2.5

[a] Sources of data. Free sugars: Somogyi and Trautner [102]; Lee et al. [67]. Others: Southgate (unpublished data) and McCance and Widdowson [73].
[b] The sugar content varies with storage—some varieties may show a tenfold increase in free sugars [117].
[c] R ≡ raffinose.
[d] Tr. ≡ trace.

Fruiting Vegetables
This category includes the vegetables that are botanically fruits, such as the tomato, marrow and cucumber. Again the solids are low but carbohydrates account for more than half of them. The free sugars follow the pattern found in fruits but slightly higher amounts of pectic substances are present. The seed coats in these vegetables are frequently lignified and the lignin concentrations in these vegetables may be appreciably higher than in most other vegetables.

Methods of Analysis
Most vegetables contain a mixture of free sugars, starch and the structural polysaccharides. A comprehensive analysis of their carbohydrates therefore requires the application of the complete analytical scheme.

Free sugars are readily extracted in aqueous alcohols and the residue can be used for the polysaccharide analysis. Care is necessary, particularly with root vegetables, to avoid hydrolysis of fructans.

FRUITS

Fruits contain a higher proportion of free sugars than vegetables and a lower proportion of unavailable carbohydrates than most vegetables [32, 67, 133].

The free sugars in fruits are usually a mixture of glucose, fructose and sucrose. Occasionally other sugars are present, as in the avocado pear. The proportions of the different sugars is reasonably characteristic of the fruit, although different varieties of the same fruit show some variation. Table 7.3 gives some examples of the composition of the free sugars in a range of fruits. Starch can frequently be detected and some fruits contain appreciable amounts, depending on the state of maturity of the sample analysed.

The cell wall constituents of fruits are derived from more delicate and less mature walls than vegetables and the amounts of pectic substances present may be as high as 1·80 g/100 g fresh fruit. Some typical values have been included in the table, together with estimates of the total unavailable carbohydrates present.

Methods of Analysis
The comprehensive analysis of the carbohydrates in fruits requires the application of a complete scheme of analysis. The free sugars are most conveniently analysed by enzymatic methods.

TABLE 7.3

TYPICAL VALUES FOR THE CARBOHYDRATES IN FRUITS[a,b] (g MONOSACCHARIDES/100 g FRESH WEIGHT)

Fruit	Number of varieties	Total solids	Glucose	Fructose	Sucrose	Other sugars	Starch	Pectin	Non-[e,f] cellulosic poly-saccharides	Cellulose	Lignin[g]	Total dietary fibre
Apples, eating	Several	14 ~ 19	1·51	6·02	3·01	M.[d]	0·30	0·38	0·95	0·47	Tr.	1·42
Apricots	Several	13 ~ 15	1·62	0·85	5·46	M.	0	0·46	—	—	—	2·10
Banana	One	29	5·82	3·78	6·58	—	3·02[c]	—	1·11	0·37	0·27	1·75
Blackberries	Several	15 ~ 18	2·67	2·48	0·47	M.	0	0·32	—	—	—	7·3
Cherries, sweet	Several	19 ~ 22	6·14	7·08	0·19	M.	0	0·09	0·92	0·25	0·07	1·24
Currants, black	Two	22	3·00	3·89	1·44	M.	0·62	0·58	—	—	—	8·7
red	Two	16 ~ 17	2·60	2·01	0·48	—	0	—	—	—	—	6·3
white	Two	17	3·08	3·00	0·58	—	0	—	—	—	—	6·8
Gooseberries, ripe	Several	14 ~ 16	3·57	3·70	1·09	—	0	0·52	—	—	—	3·5
Grapefruit	Several	9	2·05	1·75	2·29	—	0	—	0·44	0·05	0·08	0·6
Lemons	One	15	1·40	1·35	0·41	—	0	0·51	—	—	—	5·2
Oranges	Several	14	2·48	2·19	3·67	—	0	0·46	—	—	—	2·0
Peaches	Several	9 ~ 15	0·96	1·13	6·83	M.	0	0·40	1·46	0·21	0·61	2·28
Pears	Several	13 ~ 17	1·67	6·53	1·33	M.	Tr.	—	1·23	0·63	0·40	2·26
Pineapples	One	15·7	2·32	1·42	7·89	—	0	0·04	—	—	—	1·20
Plums	Several	15 ~ 23	3·84	1·74	4·37	—	Tr.	0·70	0·99	0·23	0·29	1·51
Raspberries	Two	17 ~ 20	2·15	2·00	2·02	—	Tr.	0·30	—	—	—	7·40
Strawberries	Several	8 ~ 11	2·19	2·56	1·12	—	Tr.	0·33	0·98	0·34	0·80	2·12

[a] Sources of data. Free sugars: Widdowson and McCance [133]; Dako *et al.* [32]; Lee *et al.* [67]. Others: Jacobs [58]; Kefford [61]; McCance and Widdowson [73]; Southgate (unpublished data).

[b] The sugar content of most fruits depends on the variety, the degree of maturity and the overall level of illumination the fruit has received.

[c] The starch content of the unripe banana may be as high as 20 g/100 g and the free sugars of the order of 1 g/100 g.

[d] M ≡ maltose.

[e] Includes the uronic acid components.

[f] — signifies no data available on composition of dietary fibre.

[g] Lignin as g/100 g.

The determination of pectic substances in fruit has considerable technological importance and methods are available where the pectic substances are precipitated as the calcium salts or pectic acid, which are then measured gravimetrically. More specific techniques involving enzymatic hydrolysis with polygalacturonase preparations have also been proposed. These procedures are described in the chapter on selected methods (Chapter 8).

NUTS

Nuts contain relatively low concentrations of free sugars ($2 \sim 5\%$), the starch content of many nuts is also low, with the exception of chestnuts, which may contain about 30% on a fresh basis [73, 74]. The unavailable carbohydrate contents seem to be about $6 \sim 10\%$, but only a very limited number of analyses are available.

The complete analysis of the carbohydrates in nuts thus requires the application of a detailed scheme.

SUGARS, PRESERVES AND CONFECTIONERY

This group of foods includes a wide range of food types with very different carbohydrate compositions. The general feature of these foods is a high concentration of free sugars; these free sugars may be sucrose or, more usually, a mixture of glucose, fructose and sucrose with, in many cases, the maltose components derived from glucose syrups. Starches and dextrins may also be present but usually as a minor component of the available carbohydrates.

The amounts of unavailable carbohydrates are usually low and are mostly non-cellulosic polysaccharides, either of natural origin such as the pectin in jams and marmalades or derived from food additives, which may include pectin and various gums. Modified starches are also widely used in these products for producing the desired physical texture.

These foods present considerable difficulty to the analyst because of the wide range of carbohydrate species present. A comprehensive analysis of some of these products would be very time-consuming and would require the application of the full range of techniques discussed earlier.

Simplified methods, often of an empirical kind, are widely used in the analysis of these products. These are summarised below.

Sugars and Sugar Products
These are frequently analysed by physical techniques because the products have a relatively simple carbohydrate composition. Most of the methods described by the ICUMSA have been developed for foods of this kind [128].

Preserves
In these foods the analysis is often restricted to the measurement of soluble sugars. In most products these are a mixture of invert sugar and sucrose and the standard reducing sugar methods are used [58].

Confectionery
Where only sucrose or sucrose with invert sugar is present, standard reducing sugar methods are often employed. The widespread use of glucose syrups has meant that these methods are unreliable and the use of enzymatic methods using amyloglucosidase procedures for hydrolysis is increasing. The routine quality control of these products can frequently be made using empirical physical measurements. However, these provide only approximate values for the carbohydrates present.

Choice of Procedure for Comprehensive Analysis
The free sugars can be extracted with aqueous alcohol without major difficulties. The high concentration of sugars in many of these products means that a high ratio of extracting alcohol to sample is required, and that the analyst must confirm by qualitative means that the extraction is complete.

A qualitative examination of both aqueous alcoholic and hot water extracts by paper or thin-layer chromatography is essential, as the formulation of similar products may be different. The choice of method for the analysis of the mixture of sugars present can then be made according to the alternatives set out in Fig. 3.2 (p. 46).

The residue after extraction with aqueous alcohol will still contain fat and it is usually more convenient to remove this with diethyl ether before attempting to fractionate the polysaccharides. Enzymatic hydrolysis of the starch is preferable if unavailable carbohydrates are likely to be present. A chromatographic examination of a dilute acid hydrolysate will show whether or not any interfering polysaccharides are present.

BEVERAGES

These are best considered under three categories, ready to consume non-alcoholic and alcoholic beverages and concentrated beverage powders. The first category also includes beverages that are simply diluted before consumption.

Ready-to-Consume Non-Alcoholic Beverages
The carbohydrates in these products are mainly free sugars with small amounts of soluble pectic substances in those beverages derived from fruits. The sugars present may be sucrose alone or in combination with invert sugar, or the mixture of sugars derived from glucose syrups. Preparation of the sample for analysis is usually simple and involves removal of dissolved gases and precipitation of pigments and interfering substances. Saturated lead acetate provides a convenient way of precipitating many pigments in this type of product.

The method of analysis chosen depends on the mixture of sugars present; complex mixtures are probably best analysed enzymatically but mixtures of glucose, fructose and sucrose can be measured quite satisfactorily with reducing sugar methods. Dilute acid hydrolysis of any maltose derivatives present can result in degradation of the fructose, and therefore hydrolysis with an amyloglucosidase preparation is preferable.

Alcoholic Beverages
The analysis of these follows the principles applicable to the non-alcoholic beverages.

Wines
The mixture of sugars present is usually restricted to glucose and fructose in table wines, sucrose may also be present in these wines and is usually present in sweet fortified wines. Dry wines may contain as little as 2 g/litre and fortified sweet wines 120 g/litre. The analysis involves the removal of pigments with saturated lead acetate and the measurement of glucose, fructose and sucrose in the extract. Reducing sugar methods can be used and it is necessary to remove virtually all the alcohol present by evaporation.

Beers
The analysis of these beverages is more complex as they may contain soluble polysaccharides that could be considered as dextrins and β-glucans from the cereals used in their preparation; stabilising polysaccharide additives may also be present [136].

A qualitative examination should precede analysis, the free sugars usually present are glucose, maltose and traces of higher maltose homologues; lactose may also be present in some products.

Dilute acid hydrolysis may give reasonable values for the total carbohydrate present but the values obtained by this method are usually higher than after enzymatic hydrolysis of the α-glucans alone.

Spirits and Liqueurs
Spirits usually contain virtually no carbohydrate but some liqueurs have large amounts of sugar used in their formulation. This is usually added as sucrose but some inversion may occur on standing, their analysis therefore involves the analysis of an invert sugar/sucrose mixture after removal of alcohol and dilution. Approximate values for the sugar content of liqueurs can be obtained by measuring the specific gravity of the sample after removal of the alcohol by distillation.

Concentrated Beverage Powders
These contain a wide range of carbohydrates and their analysis involves the same type of problems as the analysis of confectionery. The free sugars can include sucrose and mixtures of invert sugar and the carbohydrates from glucose syrups. Some products contain milk powder, which contributes lactose, while others contain maltodextrins, which contain a complex range of glucose polymers of different molecular size, including those insoluble in aqueous alcohol and giving a colour with iodine in potassium iodide.

A comprehensive analysis of the carbohydrates in these products must therefore be based on the principles discussed earlier (Chapter 3 (Summary)). The analysis of the free sugars must be preceded by a qualitative examination.

Simplified procedures can be used in the quality control of these products, for example measurement of reducing sugars, but these will only give approximate values for carbohydrate.

Cocoa based products appear to contain appreciable amounts of unavailable carbohydrates and significant amounts of material that analyses as lignin.

FOOD MIXTURES, PREPARED DISHES AND MIXED DIETS

In order to analyse the carbohydrates in samples of this type, the analyst must depend on a careful series of qualitative examinations at each stage of analysis, preferably coupled with a knowledge of the type of ingredients or items making up the mixture or diet.

In carrying out the analysis, the general principles of the approach to the analysis of mixtures must be followed. Given that this approach is adopted and care is taken in the interpretation of the results, reasonably reliable results can be obtained by the procedure described in Chapter 3 (Summary). At the present time, however, it is not possible to define a routine procedure that can be applied to all foods and which will produce results that do not need knowledgeable interpretation.

The preparation of samples of mixed diets or mixtures of foods can involve difficulties if isolated pieces of vegetables or fruit are present. These, particularly their skins, are difficult to homogenise in the conventional blade homogeniser and care is necessary to avoid distorting the composition of the sample.

The homogenates of mixed diets can be preserved by saturation with benzoic acid and storage at $0 \sim 4\,^{\circ}C$ if deep-freeze storage is not available.

CHAPTER 8

Selected Methods

This chapter gives details of the methods that have been mentioned in the earlier chapters. The aim has been to provide sufficient practical details of the methods to enable the intending analyst to perform these procedures adequately.

In compiling this section I have tried to describe a range of examples. However, as it is impossible to describe all the procedures that have been found of value in the analysis of food carbohydrates, some selection has been necessary.

In making this selection I have, in general, only described methods which I, or colleagues with whom I have been closely associated, have had direct experience.

This selection procedure may appear somewhat arbitrary and unfair and I apologise to all those authors who have described comparable methods.

PREPARATION OF STANDARD SOLUTIONS

Although relatively pure monosaccharides are fairly easy to obtain, as purchased they may contain water and it is usually best to dry a portion of the sugar at reduced pressure over P_2O_5 for several days before weighing out the standard amount.

Some sugars are extremely hygroscopic and it may be necessary to check the strength of the standard by polarimetry.

Standard solutions prepared at a strength of 10 g/litre in saturated benzoic acid can be used in many methods. These solutions appear to remain stable for at least 6 months.

Polysaccharides of known purity are rarely available commercially and, if standards are required, it is usually better to prepare a specimen of the polysaccharide from a defined source and to characterise this preparation directly. Methods for the preparation of examples of many polysaccharides are given in *Methods in Carbohydrate Chemistry*, Volume V (*see* Bibliography).

PHYSICAL METHODS

Measurement of Specific Gravity [4]

This method is accurate only for pure sucrose solutions but is extensively used for obtaining approximate results with many liquid sugar products containing invert sugar and other non-sucrose solids.

Using a Hydrometer

Two types are commonly used, calibrated in °Brix, which corresponds with sucrose concentrations by weight, and in Baumé Modulus (°Bé).

The hydrometer must be used in a cylinder of sufficient diameter to avoid surface tension and viscosity effects between the spindle of the hydrometer and the wall of the vessel, a clearance of at least 12 mm or larger is adequate. The temperature of measurement should ideally be that at which the hydrometer was calibrated, but tables of correction for temperature differences are available. The solution being measured should be left until all air bubbles have dispersed and fatty or waxy materials should be skimmed off. The spindle should be lowered carefully into the solution. The sucrose content is then determined by consultation of tables relating specific gravity to sucrose content.

The values on the Baumé scale are calculated according to the relation

$$°Bé = 145 - \frac{145}{\text{Spec. grav. } 20/20°C}$$

If the solution is too dense for this procedure, it is convenient to dilute the solution with an equal weight of water and multiply the °Brix by 2.

By Means of a Pycnometer

A 100 ml or 50 ml volumetric flask that has been calibrated accurately makes a useful pycnometer for sugar solutions. The specific gravity should be measured at 20/20 °C or 20/4 °C and the sucrose content read from the appropriate table.

Measurement of Refractive Index [4]

This method is accurate only for pure sucrose solutions but is also widely used for approximate values for other solutions of sugars. The solution must be clear with no undissolved solids.

The most convenient refractometers for this method are those calibrated directly in sucrose units. The refractive index measured directly on other instruments is converted to sucrose content by consulting the appropriate

tables. The manufacturers' operational instructions should be followed carefully and the temperature of measurements must be maintained at that used for calibration. Clear but dark solutions may need to be diluted before measurement.

Polarimetric Methods [4]

Polarimetric measurements are made using light of a fixed wavelength, the D-line of sodium being the standard. The specific rotation is dependent on the temperature of measurement and, for meaningful results, the rotation of solutions should be measured at a fixed temperature.

The specifications for the instruments have been laid down by the International Commission on Uniform Methods of Sugar Analysis [128], who have specified that the instruments should be calibrated in units of the International Sugar Scale.

100° on this scale is the polarisation of a normal solution of pure sucrose (26·000 g/100 ml) at 20 C in a 200 mm tube using white light and a dichromate filter.

At 5461 Å, $100°S = 40·690 \pm 0·002°$ at 20°C,
  i.e. $1° = 2·4576°S$.
At 5892 Å, $100°S = 34·620 \pm 0·002°$ at 20°C,
  i.e. $1° = 2·8885°S$.

Determination of Polarisation of Sucrose

Use $26 \pm 0·002$ g for 100 ml and bring the solution to the correct temperature. Then clarify the solution by the addition of dry basic lead acetate.* Shake the mixture and then filter, rejecting the first 25 ml of filtrate. Measure the polarisation in a 200 mm tube, unless the solution is too dark, when a 100 mm tube may be used.

Mutarotation

Many products show mutarotation when freshly prepared solutions are used. As only constant readings are of any value, the solution must be prepared overnight; alternatively the neutral (pH 7) solution may be heated to boiling or made slightly alkaline with ammonia or sodium carbonate.

Sucrose after Inversion [4]

Polarisation before and after inversion can be used as a method for measuring sucrose in the presence of other sugars.

* Precise directions for the preparation of this reagent and other clearing agents should be followed if accurate results are to be obtained.

In the absence of raffinose invertase inversion can be used.

Procedure. Dissolve double the normal weight (52 g) in water and make up to 200 ml with the necessary clarifying agent. Filter the mixture, discarding the first 25 ml of filtrate.

Dilute one 50 ml portion of the filtrate to 100 ml with water and polarise directly in a 200 mm tube. The result multiplied by two is the polarisation before inversion (P).

Treat the second portion with invertase and then dilute and polarise as before, making a correction for the optical activity of the invertase. The reading multiplied by two equals the normal inverted polarisation (I).

$$\%S = \frac{100(\text{Direct reading} - \text{Inverted reading})}{132 \cdot 1 - 0 \cdot 0833(13 - m) - 0 \cdot 53(t - 20)}$$

where $\%S = \%$ sucrose, t = temperature of polarisation ($^\circ$C), m = g total solids in inverted solution (total solids in 50 ml original).

Inversion with HCl can also be used with a slightly different formula

$$\%S = \frac{100(P - I)}{132 \cdot 56 - 0 \cdot 0794(13 - m) - 0 \cdot 53(t - 20)}$$

When inversion is made at room temperature, a slight correction of the formula is again necessary to

$$\%S = \frac{100(P - I)}{132 \cdot 66 - 0 \cdot 0794(13 - m) - 0 \cdot 53(t - 20)}$$

REDUCING SUGAR METHODS

The details of the procedures are based on the descriptions in the AOAC Official Methods [4, 128].

Lane–Eynon Method

Careful attention to the experimental details is essential if consistent results are to be obtained. The sugar solutions should be neutral or very nearly so and contain no lead or other reducing substances.

Reagents

Soxhlet Modification of Fehlings Reagent

   Copper sulphate. Dissolve 34·639 g of $CuSO_4 . 5H_2O$ in water and dilute to 500 ml. Filter through glass wool or glass filter paper if not clear. For very accurate work the copper content of the solution should be measured by electrolysis and adjusted to 440·9 mg Cu per 25 ml. In practice, the use of AnalaR $CuSO_4 . 5H_2O$ and careful weighing can make this precaution unnecessary.

Alkaline tartrate. Dissolve 173 g potassium sodium tartrate $(4H_2O)$ and 50 g NaOH in water and dilute to 500 ml. Leave for two days and filter off any precipitate through asbestos.

These two reagents are mixed in equal quantities immediately before use; they are quite stable when stored at room temperature. The cupric sulphate solution may deposit a precipitate unless it is very slightly acidified.

Standard Invert Sugar Solution

Prepare a 1 % (w/v) standard by dissolving exactly 9·5 g sucrose (AnalaR) in water, add 5 ml HCl and dilute to 100 ml. Allow the solution to invert at room temperature for several days and then dilute to 1 litre. This acidified solution is stable for several months. Alternative methods of preparing standard sugar solutions are discussed earlier.

Indicator. 0·2 % methylene blue in water.

Standardisation

Standardisation is carried out by placing 10 or 25 ml of the mixed Soxhlet reagents in a 300–400 ml conical flask. Calculate the amount of the standard sugar solution (between 15 and 50 ml) that will completely reduce the reagent and add all but 0·5–1·0 ml of it to the mixed reagents. Then heat the mixture to boiling (usually on a wire gauze over a burner) and maintain at boiling for 2 min. Then add the indicator (1 ml) without removing the heat and complete the titration by dropwise addition of the sugar solution within a total boiling time of 3 min. until the solution is completely decolorised.

Multiply the titre by the sugar concentration (mg/ml) in the standard to give the total sugar required to reduce the copper, then compare this value with Table A.2. The standardisation may show a small deviation from the tabulated values, large deviations are indicative of faulty technique or reagents.

Determination

For accurate work a two-stage procedure is necessary but if approximate values (± 1%) are required, the first incremental procedure may be sufficient.

Incremental procedure. Add 15 ml of sugar solution to 10 or 25 ml mixed Soxhlet reagents and heat to boiling; boil for about 15 s then rapidly add a further amount of sugar solution until the faintest blue tinge remains. Add the indicator and continue titration dropwise.

Standard method. Repeat the determination, adding almost the entire volume required to effect reduction then proceed as in the standardisation procedure.

Munson and Walker Method

Careful attention to the experimental details is essential; the sugar solutions should be neutral or faintly alkaline and free from lead and other reducing substances.

Reagents

Soxhlet Modification of Fehlings Reagent

Standard invert sugar. As in Lane–Eynon method (*see* above).

Asbestos. The asbestos should be of the amphibole variety and digested with HCl $(1 + 3)$ for 2–3 days. Wash free from acid and digest for a similar time with 10% (w/v) NaOH and then for a few hours with the alkaline tartrate reagent. Then wash the asbestos with water, digest with HNO_3 $(1 + 3)$ and wash with water again.

Finally, shake with water to give a fine pulp. The Gooch crucibles are prepared with a filter mat about $\frac{1}{4}$ in thick and washed with water, followed by alcohol and ether. Dry the crucibles for 30 min at 100 C, cool in a desiccator and weigh.

Determination

The standard procedure for precipitation of the cuprous oxide is as follows—several different techniques can be used for measuring the cuprous oxide.

Pipette 25 ml each of the $CuSO_4$ and alkaline tartrate reagents into a 400 ml beaker and add up to 50 ml of the sugar solution. If a smaller volume is used, water should be added to make the final volume 100 ml.

Then heat the beaker over a Bunsen burner on an asbestos gauze, regulating the flame so that boiling begins in 4 min and continues boiling for exactly 2 min. These times are critical and variations of more than ± 15 s will affect the results obtained. Cover the beaker with a watch glass during the heating.

Filter the hot solution at once through the asbestos mat using suction. Wash the cuprous oxide precipitate throroughly with water at 60 C and either weigh directly or measure by titration or electrolysis.

A blank determination should be carried out using 50 ml mixed reagents and 50 ml water. If the weight of Cu_2O exceeds 0·5 mg, results should be corrected accordingly. The alkaline tartrate reagent slowly deteriorates on standing and the blank values increase as the reagent ages.

Measurement of Reduced Copper
(a) By direct weighing: collect the Cu_2O as directed above and wash the precipitate with 10 ml alcohol followed by 10 ml ether. Dry for 30 min in an oven at 100°C, cool in a desiccator and then weigh.

Read off the weight of reducing sugar equivalent to the weight of Cu_2O in Table A.3.

The weight of cuprous oxide depends on the reducing sugar present, and whether or not sucrose is present.

(b) By titration with potassium permanganate.

Reagents
Potassium permanganate. Approximately 0·1573N containing 4·98 g/litre. Standardised against sodium oxalate. 1 ml = 10 mg Cu.

Ferric sulphate. Dissolve 135 g $FeNH_4(SO_4)_2 . 12H_2O$ or 55 g anhydrous $Fe_2(SO_4)_3$ in water and dilute to 1 litre. Measure blank by titration of 50 ml acidified with 20 ml $4N–H_2SO_4$.

Ferrous phenanthroline indicator. Dissolve 0·7425 g *o*-phenanthroline H_2O in 25 ml 0·025M–$FeSO_4$ (6·95 g $FeSO_4 . 7H_2O$/litre).

Determination
Filter the Cu_2O through the crucible, washing the beaker and precipitate thoroughly. Transfer the asbestos pad to the beaker and add 50 ml ferric sulphate solution. Stir vigorously until the cuprous oxide precipitate is completely dissolved.

Add 20 ml $4N\ H_2SO_4$ and titrate with standard $KMnO_4$. As the end-point approaches add one drop of indicator; at the end-point the brownish colour changes to green.

Read off the weight of reducing sugar equivalent to the weight of copper from Table A.4.

The copper can also be measured by titration with thiosulphate or electrolytically by deposition from nitric acid.

Nelson–Somogyi Micro-colorimetric Method
This is the modification used by Somogyi [101] of a colorimetric method for the measurement of reducing sugars.

Reagents

Copper sulphate reagent. Dissolve 28 g anhydrous Na_2HPO_4 and 4 g sodium potassium tartrate in about 700 ml water, add 100 ml N NaOH with stirring and then 80 ml 10 % (w/v) cupric sulphate. Add 180 g anhydrous Na_2SO_4 when solution is complete and dilute to 1 litre. Leave for a day, then decant the clear supernatant. This reagent keeps indefinitely.

Arsenomolybdate reagent. Dissolve 25 g ammonium molybdate in 450 ml water, add 21 ml H_2SO_4 (conc.) and mix; then add 3 g $Na_2HAsO_4 . 7H_2O$ dissolved in 25 ml water. Mix and incubate at 37 $^\circ$C for 24–48 h. Store in a brown bottle, preferably in a cupboard.

Standard glucose. A stock standard 1 % (w/v) glucose solution in saturated benzoic acid is diluted to give working standards containing 50, 150 and 300 μg/ml.

Determination

A blank and a series of standards must be carried through with each series of unknown samples. The reaction is carried out in glass tubes (16 mm × 150 mm, calibrated between 10 ml and 25 ml) either with glass stoppers, or covered with a glass bulb (or marble). Measure 2 ml of copper reagent and 2 ml of test solution into each tube and place the tubes in a boiling water-bath for 10 min and then cool for 5 min in running water. Add 1 ml of the arsenomolybdate reagent and, after mixing, dilute the contents of the tube to a definite volume between 10 and 25 ml, depending on the colour density.

Measure the absorbance at 500 or 520 nm against the blank (the maximum absorbance is at 660 nm).

Notes

(1) The colour is very stable.

(2) The heating conditions must be rigidly standardised and all the tubes in a series should be placed in the water-bath, removed and cooled simultaneously. The boiling water-bath should be heated in such a way that it does not go off the boil for more than a few seconds when the tubes are added to it.

Alkaline Ferricyanide Method [130]

Many versions of this method have been used and several automated versions have been described. The method is based on the principle that above pH 10·5 sugars reduce ferricyanide to ferrocyanide, which reacts with ferric ions to produce Prussian blue.

Reagents
  Alkaline cyanide. 0·53% (w/v) sodium carbonate containing 0·065%(w/v) potassium cyanide.
  Potassium ferricyanide. 0·05% (w/v) solution in water.
  Ferric ammonium sulphate. 1·5 g dissolved in 1 litre of 0·05 N H_2SO_4.

Method
Mix 1–3 ml of reducing sugar solution (containing between 1 and 9 μg reducing sugar) with 1 ml of alkaline cyanide and 1 ml of ferricyanide solution. Heat the mixture for 15 min in a boiling water-bath. A blue colour appears; this is stable and can be measured at 700 nm. The reaction obeys Beer's law but it is usual to carry standard amounts of sugar through with each series of unknowns.

Comments
The AOAC official methods describe a version of this method where a volumetric titration with thiosulphate is used to measure the reduction and Technicon describe an automated version.
  Standardisation of heating conditions is essential for the best results.

Reaction with Triphenyl Tetrazolium [130]
Reducing sugars react with 2,3,5 triphenyl tetrazolium to give triphenylformazon, which is soluble in organic solvents and produces a cherry red colour.

Reagents
These are as follows: 2, 3, 5 Triphenyl tetrazolium, 0·3% (w/v) aqueous solution; 2N NaOH; 2·1N Acetic acid.

Determination
Add 1 ml of tetrazolium reagent and 1 ml of 2N NaOH to 2 ml of sugar solution (20–200 μg glucose). Heat the mixture in a boiling water-bath for 3 min and immediately acidify with 1 ml of 2·1N acetic acid. Dilute the mixture with iso-propanol or ethanol to 25 ml.
  The colour has an absorbance maximum at 485 nm and the colour is linear over the range 10–100 μg.

CONDENSATION REACTIONS

Many of these reactions have been described in the literature and have been reviewed by Dische [35] (and in [130]). A limited number of these reactions have come into widespread use and the more important of these are described below.

Phenol/Sulphuric Reaction [130]

Virtually all classes of carbohydrate react with this reagent and produce a stable colour. The strong acid reagent will react with all polysaccharide dusts and it is important to exclude the possibility of contaminating the solutions or apparatus with cellulose fibres from paper, etc.

Reagents

Phenol. Redistilled phenol (reagent grade); 50 g dissolved in water and diluted to 1 litre.

Sulphuric acid. Concentrated (AnalaR). Some batches of acid will give elevated blank values and anomalous colours. This can usually be avoided if microanalytical grade reagents are used.

Determination

Blanks and standard solutions must be carried through with each batch of unknowns. Pipette 1 ml of the aqueous sugar solution (10–70 µg) into a tube, add 1 ml of phenol solution and mix. Add 5 ml H_2SO_4 from a fast-flowing pipette to the tube and mix the contents rapidly. The tube should be agitated during addition of the acid and a uniform procedure for all the tubes in a series is essential. After 10 min shake the tubes again and place them in a water-bath at 25–30 °C for 20 min. The colour produced is stable for several hours and has maxima of 490 nm for hexoses and methylated hexoses, and 480 nm for pentoses, uronic acids and their methylated derivatives.

Anthrone Reaction

Most carbohydrates react to some extent but under the conditions described the reaction provides a reasonably specific method for hexoses. All polysaccharides will react in the strong acid and contamination with cellulose dust or fibres must be rigorously excluded. The earlier descriptions of the method [35] involved the use of the reagent in concentrated sulphuric acid but this is very unstable.

Variation in the blank can be very troublesome and recrystallisation of the anthrone is necessary if low and acceptable blanks are to be obtained.

As with all condensation reactions, the conditions of heating and cooling must be closely standardised and all the tubes in a series should be treated simultaneously at the heating and cooling stages. The method described is that of Roe [96].

Reagents

Anthrone/thiourea. Stock 66 % (v/v) H_2SO_4 is prepared by adding 660 ml H_2SO_4 cautiously with stirring and external cooling to 340 ml water in a large beaker. Prepare several litres at a time and leave to cool. Dissolve 10 g thiourea and 0·5 g anthrone (9,10 dihydro-9-oxoanthracene) in 1 litre of this acid by warming the mixture to 80–90 ˙C. Store at 0–4 ˙C. The colour of the reagent increases slowly with time and the colour yields tend to decline after 2 weeks.

Standard glucose. A stock standard is diluted to give standards in the range 25–200 μg/ml.

Determination

Blank and standard amounts of glucose should be carried through with each series of unknowns. Pipette 1 ml of the test solution into a glass stoppered tube and add 10 ml of anthrone reagent. Swirl to mix the contents and stopper the tube firmly. Place in a water-bath at room temperature to bring to equilibrium and then into a boiling water-bath for 15 min. Cool to room temperature in a tap-water bath and leave in the dark.

Measure the absorbance at 620 nm after 20–30 min.

Comments

This procedure is technically easier than the original Dische method [35] by virtue of the lower strength of acid used. The method is less sensitive but this is not usually a problem in food analysis. The method will tolerate modest concentrations of ethanol or methanol (up to say 5 % (w/v)) but for the most accurate work the standards used to construct the standard curve should also contain alcohol at the same strength as the unknowns.

Several modifications of the heating routine have been suggested, and these permit a semi-quantitative analysis of mixtures of hexoses. These require very close control of the initial mixing of reagent and sugar solution and of the subsequent heating in the water-bath. Fructose and fructose-containing polysaccharides will react quite readily at the temperature produced by mixing the sample and reagents, and can be measured without further heating if suitable standards are used.

Resorcinol Method (Ketoses)

Resorcinol reacts with sugars in strong acid to give coloured complexes and, although all hexoses react, the colour yield with ketoses is very much greater than that with aldoses—so that the method can be used to give values for ketohexoses. The procedure described is that of Kulka [65].

Reagents
   Resorcinol. 0·5% (w/v; AnalaR) resorcinol in absolute ethanol. This reagent is reasonably stable if kept in the dark.
   HCl/ferric ammonium sulphate. Conc. HCl containing 0·216 g ferric ammonium sulphate per litre.

Determination
Pipette 2 ml of sample (5–50 μg fructose/ml) into a stoppered tube and add 3 ml resorcinol reagent followed by 3 ml of the acid. Heat for 40 min in a water-bath at 80 °C ± 0·5 deg C; cool in ice-water for 1·5 min and measure the absorbance at 480 nm.
   The colour is stable for 5 h, provided it is not exposed to bright light.

Orcinol/Ferric Chloride Reaction (Pentoses)
Orcinol reacts with carbohydrates in strong acid. In sulphuric acid most carbohydrates react to give a coloured product and the reaction has been used to give a method for total carbohydrate. The procedure described is a modification in which the colour yield from pentose sugars is very much higher than the yield from hexoses. It is based on the method of Mejbaum as modified by Albaum and Umbreit [1] in which arabinose and xylose give equal colour yields.

Reagents
   Orcinol. 10% (w/v) solution in ethanol.
   Ferric chloride/HCl. 0·1% (w/v) ferric chloride (anhydrous) in concentrated HCl (AnalaR).
   These two reagents are mixed in the proportion of 1 plus 10 (orcinol + acid) immediately before use.
   Standard pentose. A stock solution of arabinose (1% (w/v) in saturated benzoic acid) is diluted to give 2, 5 and 10 μg/ml working standards.

Determination
Blank and standard amounts of pentose should be carried through with each series of unknowns. The reaction is conveniently carried out in glass stoppered tubes.
   Pipette 3 ml of test solution into the tube and add 3 ml of

orcinol/$FeCl_3$/HCl reagent. Mix and place in a boiling water-bath for 45 min. Cool and measure the absorbance at 670 nm. The colour is reasonably stable.

Comments

Standardisation of heating is essential. Hexoses give about 5 % of the colour yield of pentoses, and the pentose values should be corrected for the hexose sugars present on this basis.

Carbazole Reaction (Uronic Acids)

The original Dische procedure for carbazole has also been the subject of many modifications and Dische [35] has described a number of variants that permit the qualitative and partially quantitative measurement of different uronic acids. The procedure described is that of Bitter and Muir [16].

Reagents

Borate/H_2SO_4. 0·025M sodium tetraborate . $10H_2O$ in concentrated H_2SO_4 (AnalaR).

Carbazole. 0·125 % (w/v) in absolute ethanol. This reagent should be stored in the dark at 0–4 C, where it is stable for about 6 months.

Standard uronic acid. Galacturonolactone is a convenient standard. The stock (1 % (w/v) in saturated benzoic acid) is diluted to give standards with 4–40 μg/ml.

Determination

Blank and standards must be carried through with every series of unknowns. Pipette 5 ml of borate/H_2SO_4 into a stoppered tube and cool to 4 C. Add 1 ml of test solution and gently mix. Heat the tube in a boiling water-bath for 10 min and cool to room temperature.

Add 0·2 ml of carbazole reagent and mix the contents of the tube. Measure the absorbance at 530 nm. The colour is stable for about 16 h.

Comments

Extraneous colours can be generated by chloride ions and many organic substances will give a non-specific brown colour with the acid. In these cases it is necessary to run a series of samples with and without carbazole.

Galacturonic and glucuronic acids give identical colour yields but iduronic acid gives only 83 % of the colour yield.

Several automated versions of the carbazole procedure are available, for example Balazs *et al.* [10].

DECARBOXYLATING REACTIONS (URONIC ACIDS) [25, 30]

Uronic acids in hot 12 % (v/v) HCl undergo decarboxylation and, if the conditions are controlled, the carbon dioxide evolved can be stoichiometrically related to the amount of uronic acid present. An apparatus with an absorption train for liberated carbon dioxide is described by Whistler and Feather [129]. The principle is that a carrier gas (nitrogen) is passed into a reaction flask in which the sample is refluxed with the HCl. The gas is then passed into a soda lime absorption tube.

The temperature of reflux is about 130 °C and, in the apparatus described, a sample sufficient to generate 40 mg of carbon dioxide is required. A semi-micro-manometric method has been described [116] but in general the sensitivity of these procedures is not particularly high.

BIOCHEMICAL PROCEDURES

In this section analytical procedures based on enzymatic procedures have been grouped together [13, 20].

Glucose Oxidase Procedure

This is a specific and sensitive method for glucose based on the principle that the glucose is oxidised with glucose oxidase (E.C.1.1.34) to form peroxide, and this reacts with a dye in the presence of peroxidase [11].

Reagents

Glucose oxidase. A range of preparations of this enzyme are available.
Peroxidase. Usually a horse radish preparation.
Chromogen. o-Dianisidine . 2HCl. This reagent should be handled with caution and alternative reagents are being developed, *e.g.* 2,2′ Azino-di-[3-ethyl-benzthiazoline sulphonate 6].
Acetate buffer. pH 5·5, 0·1M. Dissolve 13·608 g sodium acetate . $3H_2O$ and dilute to 1 litre, add 2·7 ml of acetic acid and adjust the pH as necessary. Dissolve 40 mg chromogen, 40 mg peroxidase and 0·4 ml of oxidase (≡ 400 units) in 0·1M acetate and dilute to 100 ml with acetate buffer.
Glucose standards. 1 mg/ml. Mix solution and leave for at least 2 h to complete mutarotation. Prepare fresh on day of use.

Determination
Prepare a standard curve by pipetting 1, 2, 3 and 4 ml glucose standard into 50 ml flasks and diluting to volume. Place 2 ml of each of these standards in stoppered tubes. This is equivalent to 40, 80, 120 and 160 μg glucose.

At time zero add 1 ml test solution (oxidase + peroxidase + chromogen in buffer) to the first tube, allowing 30–60 s between additions to each tube. Mix the tubes and allow to react for exactly 30 min at 30 ° C. Immediately stop the reaction by pipetting 10 ml H_2SO_4 (1 + 3) into each tube. Mix, cool to room temperature and measure against a reagent blank at 540 nm.

Treat the sample solutions in the same way.

Comments
This procedure is very specific for glucose. Occasionally some β-fructosidase activity can be detected in the oxidase and this would lead to high results for glucose in the presence of sucrose.

Glucose and Fructose by Hexokinase Procedure
This procedure can be used for the measurement of glucose alone or glucose and fructose separately in a mixture.

The reaction is catalysed by hexokinase and leads to the production of the respective 6-phosphates in the presence of adenosine triphosphate.

$$\text{Glucose} + \text{ATP} \xrightarrow{\text{hexokinase}} \text{G-6-P} + \text{ADP}$$

$$\text{Fructose} + \text{ATP} \longrightarrow \text{F-6-P} + \text{ADP}$$

The enzyme glucose-6-phosphate dehydrogenase oxidises the glucose-6-phosphate in the presence of nicotinamide adenine dinucleotide (NADP) to 6-phosphogluconate and reduced NADP (NADPH) is formed. The amount of NADPH formed is equivalent to the amount of G-6-P and glucose, and can be measured by changes in the optical density at 340 nm and 366 nm.

On completion of the reaction, F-6-P is converted to glucose-6-phosphate with phosphoglucose isomerase, which leads to the production of more NADPH proportional to the fructose. Removal of protein is essential in the preparation of the extract, and N perchloric acid added in the proportion of 1:1 is suitable for this purpose.

Reagents
   Triethanolamine buffer. Dissolve 11·2 g triethanolamine . HCl and 0·2 g $MgSO_4$. $7H_2O$ in 150 ml water, adjust to pH 7·6 with approximately 4 ml 5N NaOH then make up to 200 ml.

NADP. Dissolve 50 mg NADP(Na$_2$) in 5 ml water. (Store at + 4 °C; stable for 4 weeks.)

ATP. Dissolve 250 mg ATP(Na$_2$) and 250 mg NaHCO$_3$ in 5 ml water. (Store at + 4 °C; stable for 4 weeks.)

Hexokinase: glucose-6-phosphodehydrogenase. Mix 0·5 ml hexokinase solution (2 mg/ml) with 0·5 ml G-6-PDH solution (1 mg/ml). (Store at + 4 °C; stable for 1 year.)

Phosphoglucose isomerase. 2 mg/ml. (Store at + 4 °C; stable for 1 year.)

Determination

The solution should contain between 0·4 and 0·8 g glucose + fructose per litre; if stronger solutions are to be analysed they should be diluted accordingly. If the sample contains a large excess of glucose reduce this by treatment with glucose oxidase before fructose is measured.

Pipette 3 ml of triethanolamine buffer at 20–25 °C into a cuvette and add 0·10 ml of NADP solution, followed by 0·10 ml ATP solution and 0·20 ml of sample. Mix the contents and measure the optical density at 340 or 366 nm (E_1). Then add the suspension of hexokinase (0·02 ml) and, after mixing, allow the reaction to go to completion for 10–15 min, then measure the optical density again (E_2). If the reaction continues after 15 min, read the optical density at 5 min intervals until it is complete. Add phosphoglucose isomerase solution (0·02 ml) and mix, read the optical density after 10 min (E_3).

$$E_2 - E_1 = \Delta E \text{ glucose}$$

$$E_3 - E_2 = \Delta E \text{ fructose}$$

Calculation

The volume of the mixture for measuring glucose is 3·42 ml and for fructose 3·44 ml, and with a sample volume of 0·2 ml, at 340 nm,

$$\Delta E \text{ glucose} \times 0·495F = \text{g glucose/litre}$$

$$\Delta E \text{ fructose} \times 0·497F = \text{g fructose/litre}$$

and at 366 nm

$$\Delta E \text{ glucose} \times 0·933F = \text{g glucose/litre}$$

$$\Delta E \text{ fructose} \times 0·938F = \text{g fructose/litre}$$

where F is the dilution factor of the sample.

Sucrose by Invertase Hydrolysis

Sucrose is hydrolysed using β-fructosidase (EC.3.2.1.26; invertase) and the glucose that is produced is measured by the hexokinase procedure. Free glucose must be measured in the solution without hydrolysis and the values corrected accordingly.

Reagents

In addition to those required for a glucose estimation, the following are necessary.

Acetate buffer. Dilute 6·7 ml N NaOH and 13·5 ml N acetic acid with 180 ml water. The pH should be 4·6.

β-Fructosidase. Dissolve 10 mg in 2 ml water. (Store at + 4 °C; stable for 1 week.)

Determination

Deproteinise the solution if necessary with ice-cold N perchloric acid; centrifuge and immediately neutralise the supernatant with 2N KOH. Leave for 20 min in an ice-bath and filter off the precipitated $KClO_4$.

The solution should contain 0·4 ~ 0·8 g glucose + sucrose/litre; stronger solutions must be diluted. Remove excessive concentrations of glucose by treatment with glucose oxidase.

Mix 0·20 ml of sample with 0·50 ml acetate buffer and 0·02 ml β-fructosidase solution and leave it for 15 min at 20–25 °C. Then add 2·50 ml of ethanolamine buffer (glucose method), followed by 0·10 ml of the NADP solution and 0·10 ml of the ATP solution. After mixing, measure the optical density (E_1). Then add the hexokinase solution and leave the reaction mixture for 10–15 min before measuring the optical density again (E_2). If the reaction is not complete at this time, measure the optical density at 5-min intervals until the increase per 5 min is constant. Extrapolate E_2 to the time of addition of the hexokinase.

Calculation

[(Sucrose + glucose) − free glucose] × 2 × 0·95 = g sucrose/litre, so that with a total volume of 3·44 ml for the sucrose reaction and 3·42 ml for the glucose reaction, and with a sample volume of 0·20 ml,

at 340 nm

$$[\Delta_{A(sucrose\ test)} \times 0·497 - \Delta_{B(free\ glucose)} \times 0·495] \times 2 \times 0·95$$
$$= g\ sucrose/litre$$

at 366 nm

$$[\Delta_{sucrose} \times 0·939 - \Delta_{glucose} \times 0·933] \times 2 \times 0·95 = g\ sucrose/litre$$

Maltose

Maltose is hydrolysed by α-glucosidase (EC.3.2.1.21) to give two molecules of glucose and these are measured by the hexokinase procedure.

The reaction will measure the free glucose present and it is therefore necessary to measure free glucose in the test solution without incubation. The concentration of glucose + maltose should not be greater than 0·4 g/litre; stronger solutions must be diluted.

Reagents

Acetate buffer. 1·36 g sodium acetate . $3H_2O$ in 80 ml water, adjust pH to 6·6 with 0·1 N acetic acid and make up to 100 ml.

α-Glucosidase. 5 mg/ml (stable for 6 months at +4 °C) and reagents for the glucose by hexokinase method.

Determination

Pipette 0·5 ml acetate buffer (at 20–25 °C) into a 10 ml centrifuge tube, add 0·5 ml sample followed by 0·05 ml α-glucosidase. Mix and leave at room temperature for 10 min, then heat for 3 min in a boiling water-bath. Filter or centrifuge the mixture, then add acetic acid buffer, NADP and ATP to the supernatant. Measure the optical density at 340 or 360 nm (E_1). Add the hexokinase solution and allow the reaction to continue for 10–15 min. Measure the optical density again (E_2). If this is still changing, repeat the measurements at 5 min intervals until the rate of increase is constant. Then extrapolate E_2 to the time of addition of the hexokinase.

Calculation

Maltose = (glucose − free glucose) × 0·95 if results are to be expressed as maltose.

Lactose

Galactose is measured by following the reaction

$$\text{galactose} + NAD^+ \rightarrow \text{galactonolactone} + NADH + H^+$$

which takes place in the presence of galactose dehydrogenase. The amount of NADH formed is proportional to the galactose present.

Lactose is measured after this reaction has gone to completion by hydrolysing lactose to glucose and galactose with β-galactosidase (EC.3.2.1.23).

Reagents

   Phosphate buffer. Dissolve 4.80 g Na_2HPO_4, 0.86 g $NaH_2PO_4H_2O$ and 0.20 g $MgSO_4.7H_2O$ in 200 ml water and adjust the pH to 7.5.
   NAD. 10 mg/ml in water.
   Galactose dehydrogenase. 5 mg/ml.
   β-Galactosidase. 5 mg/ml.

Determination
The test solution should be clear and samples containing protein or fat should be deproteinised by the addition of an equal volume of N perchloric acid and the solution neutralised with 2N KOH. The solution should not contain more than 0.2–0.4 g galactose + lactose/litre; stronger solutions should be diluted accordingly.
   Pipette 3 ml of phosphate buffer and 0.10 ml of NAD solution into a cuvette, followed by 2 ml of sample. Mix the contents and determine E_1. Add the galactose dehydrogenase and determine E_2 after 30 min.

$$E_2 - E_1 = \Delta E_{Gal}$$

Then add β-galactosidase (0.02 ml) and determine E_3 after 30 min.

$$E_3 - E_2 = \Delta E_{Lac}$$

Calculation
For readings at 340 nm

$$\Delta E_{Gal} \times 0.479 \times F = \text{g galactose/litre}$$
$$\Delta E_{Lac} \times 0.915 \times F = \text{g lactose/litre}$$

and for readings at 366 nm

$$\Delta E_{Gal} \times 0.905 \times F = \text{g galactose/litre}$$
$$\Delta E_{Lac} \times 1.731 \times F = \text{g lactose/litre}$$

where F equals the dilution factor.
   Note. L-Arabinose is also oxidised in the presence of galactose dehydrogenase, so this method includes L-arabinose if it is present in the sample.

QUALITATIVE PROCEDURES

Paper Chromatography
Detailed accounts of techniques are available in a number of texts [33, 55, 69], and these should be consulted.

Solvent Mixtures

Lederer and Lederer [66] give details for the mobilities of sugars in a wide range of solvent mixtures. Three typical and widely used mixtures are 1-butanol/ethanol/water (40:11:19, by vol.); ethyl acetate/acetic acid/water (3:3:1, by vol.); and ethyl acetate/pyridine/water (10:4:3, by vol.). The mobilities expressed relative to glucose (R_{Glu}) values are shown in Table 8.1. The overall mobility in butanol systems is low. With the ethyl acetate systems separation can be improved by allowing the solvent to run off the paper.

TABLE 8.1
MOBILITY RELATIVE TO GLUCOSE IN THREE SOLVENT SYSTEMS

	Butanol/ethanol/ water (40:11:19)	*Ethyl acetate/ acetic acid/ water* (3:3:1)	*Ethyl acetate/ pyridine/water* (10:4:3)
Arabinose	1·33	1·36	1·14
Fructose	1·33	1·30	1·18
Galactose	0·78	0·95	0·82
Glucose	1·00	1·00	1·00
Mannose	1·22	1·20	1·05
Xylose	1·67	1·63	1·30
Galacturonic acid	0·22[a]	0·83	0·11
Glucuronic acid	0·01[a]	0·68	0·14
Maltose	0·02	—	0·65
Sucrose	0·03	0·67	0·72

Recalculated from Lederer and Lederer [66] and other sources.
[a] Spot streaked.

Technique

In general, descending techniques are preferable and care should be taken with the level of the solvent trough and the configuration of the paper.

The solutions should be applied to the paper with micropipettes and an effort made to conduct the method in a quantitative fashion.

Spray Reagents

Ammoniacal silver nitrate. 0·1M $AgNO_3$ in 5M NH_4OH. The chromatogram is heated for 5–30 min after spraying. Sugars are visible as brown–black spots. This reagent is rather unspecific but it is useful because so many substances react and by virtue of its sensitivity.

Aniline hydrogen phthalate. Dissolve 0·93 g aniline and 1·6 g phthalic acid in 100 ml of 1-butanol saturated with water.

Heat the chromatogram for 5 min at 100 °C after spraying. This reagent is specific for reducing sugars and many sugars, notably fructose and uronic acids, produce compounds with this reagent that fluoresce in the u.v.

Naphthoresorcinol. 0·1 % solution in 0·4N HCl in 80 % (v/v) ethanol. Trichloracetic acid, and phosphoric acid have also been used. Heat the sprayed paper for 5 min at 105 °C. Most sugars give blue spots, fructose gives a reddish spot.

Thin-layer Chromatography

The general techniques of thin-layer chromatography are well-described by Stahl [110], and the methods described here have been found to give good reliable separation.

Coating Materials

The capacity of the conventional Kieselguhr-G layers for sugars is limited and separations on these layers are unsatisfactory. Better separations are obtainable on plates that have been impregnated with borate and on plates with a cellulose layer.

Solvent Systems

As with paper chromatography, several solvent systems have been suggested. However, xylose, glucose, arabinose and mannose have very similar mobilities in most solvent systems. In general, it is necessary to run the plates more than once to achieve satisfactory resolution.

The best separations are obtained with ethyl acetate/pyridine/water (25:35:100 by vol.) on cellulose plates with multiple development of the plates. Care must be taken not to overload the layers on Kieselguhr-G. About 2 μl of a 1 % sugar solution is near optimal. On cellulose layers one can use 5 μl.

Spray Reagents

Most of the reagents used for paper chromatography can be used.

Quantitation

The quantitation of thin-layer chromatograms is not as easy as with paper chromatograms, and the technique appears to have no great value in food analysis.

ION-EXCHANGE SEPARATION OF SUGAR MIXTURES

Several procedures have been described for the separation of mixtures of sugars on ion-exchange resins, usually in the borate form. The procedure described is that of Floridi [41], which permits the satisfactory separation and measurement of very complex mixtures in about 10 h using automated analytical equipment. However, it is probable that high-pressure chromatographic separations are likely to supersede this method of separation.

Apparatus
Using Technicon equipment the following modules are required.
   (a)   A micro pump for delivering buffers to the column.
   (b)   A chromatographic column 0·6 cm × 110 cm or 0·6 cm × 140 cm of borosilicate glass with a temperature-controlled jacket. The columns are maintained at 55 °C.
   (c)   Proportioning pump for mixing the sample with reagents.
   (d)   A heating bath with a reaction coil 1·7 mm × 12 m held at 97 °C.
   (e)   A colorimeter with a 15 mm flow-cell and 420 nm interference filter.
   (f)   A single channel recorder.

Reagents
   Ion-exchange resin. Various resins were examined and Dowex 1 × 4 (Dow Chemical Co.) 200–400 mesh gave the best resolution.
   The resin was fractionated by repeated settling in 50 % (v/v) aqueous ethanol. The finest particles were used after treatment with 2N NaOH to remove chloride ions, washing with water to remove excess alkali and then treated with 0·5M boric acid until the pH was between 4·2 and 4·3. It was then equilibrated with buffer.
   The column was packed in sections while pumping buffer.
   Buffer. Three systems of elution were used. In the first a single buffer (buffer A) was used, which gave separation of two disaccharides (maltose and sucrose) and seven monosaccharides. A two-buffer system used in stepwise fashion with buffers A and B gave a good separation of all free sugars likely to be found in foods. A third system using three buffers gave a separation of a 17-component sugar mixture. The second system is described below, as this is adequate for most food analysts.
   Buffer A. 0·11M potassium tetraborate, 0·17M boric acid, pH 8·80.
   Buffer B. 0·0255M potassium tetraborate, 0·125M boric acid, pH 8·40 (adjusted with M KOH).
   Buffer C. 0·053M potassium tetraborate, 0·088M boric acid, pH 8·80.

Measurement of sugars

An orcinol/sulphuric acid method is used. The reagent is mixed with sample in the proportions of 0·035:0·065 (sample:orcinol) and segmented with air. Both the heights and the areas of the peaks are closely correlated with the amounts of the sugars applied to the column; the colour yield varies from sugar to sugar and calibration is essential.

Reagent

Orcinol. 1 g dissolved in 1 litre of 70 % (v/v) H_2SO_4. The solution is filtered before use through a sintered glass filter and stored in a brown glass bottle.

Application of sample and elution. The column is equilibrated with buffer A and the sample (containing 5 $\sim$ 50 μg of each sugar) applied in the buffer. Elution is started with buffer B at a flow-rate of 45 ml/h, after 90 min elution is continued with buffer A at 60 ml/h. This gives good resolution of 12 sugars in 336 min. Regeneration of the column is not necessary.

GAS–LIQUID CHROMATOGRAPHY

Preparation of Derivatives for Gas–Liquid Chromatography

Two classes of derivatives seem to provide the most satisfactory separation of sugar mixtures of the type likely to be encountered in food analysis; the trimethylsilyl derivatives of the sugars themselves and the alditol acetates prepared after reducing the sugars to the alditols with borohydride.

Trimethylsilyl Derivatives [9]

These were amongst the first derivatives to be used in the quantitative separation of sugar mixtures. The reaction is usually very rapid but the silylating reagents are very sensitive to water and it is usually necessary to carry out the reaction in a dry solvent (anhydrous pyridine is the most suitable for sugars).

The derivatives have a high degree of thermal stability. The main problem with quantitative separation is that for each monosaccharide four trimethylsilyl derivatives of the anomeric forms are possible and the chromatograms are usually complex.

Procedure 1

   Reagents

   Hexamethyl disilazane

   Trimethyl chlorosilane

   Pyridine (anhydrous, dried over KOH)

Method. To a sample of up to 10 mg of carbohydrate add 1·0 ml of anhydrous pyridine, 0·2 ml of hexamethyl disilazane and 0·1 ml of trimethyl chlorosilane in a plastic-stoppered glass vial. Shake the mixture for about 30 s and leave for 5 min or longer at room temperature until derivatisation is complete. Carbohydrates that dissolve slowly in the mixture may be warmed for 2–3 min at 75–85 °C.

Between 0·1 and 0·5 μl of the reaction mixture is injected onto the column.

Procedure 2 (A method developed for starch hydrolysates.)

Reagents
Hexamethyl disilazane
Trifluoroacetic acid
Pyridine
Method. Place 60–70 mg of the sugar syrup (80 % solids) in a stoppered vial and add 1 ml pyridine. When the sugars are in solution, add 0·9 ml of hexamethyl disilazane and 0·1 ml of trifluoroacetic acid; shake vigorously for 30 s then leave for 15 min with occasional shaking. Inject directly. The reaction mixture evolves heat and ammonia gas, so caution is necessary.

Alditol acetates [29]
The sugars are reduced to the alditols with borohydride and the excess boric acid removed by evaporation from methanol. The alditols are acetylated in pyridine.

Procedure 1
Reagents
Sodium borohydride
Acetic acid
Acetic anhydride
Methanol
Pyridine
Methylene chloride
Method. Dissolve 200 mg monosaccharide mixture, including 2-deoxy glucose as an internal standard, in 10 ml of water and add 400 mg sodium borohydride (with the 10 ml water). Allow the mixture to react for 1 h at room temperature; then add acetic acid until evolution of hydrogen ceases and evaporate the mixture to dryness. Add methanol (20 ml) and evaporate the solution to dryness. Reflux the residue overnight in a 1:1 mixture of acetic anhydride and pyridine. Evaporate the mixture to a syrup, add 2 ml of water and again evaporate to dryness.

Dissolve the mixture in 10 ml of methylene chloride and inject onto the column.

Procedure 2 [3]

Reduce the sugar mixture (20 mg), with myo-inositol (1 mg) as an internal standard, with 10 mg sodium borohydride in N ammonia (0·5 ml); after 1 h at room temperature add acetic acid dropwise to decompose the excess borohydride. Then evaporate the mixture to dryness after the addition of 1 ml of methanol, this evaporation being repeated five times in all.

Acetylate the alditols in sealed tubes for three hours at 121 °C with 1 ml acetic anhydride. This mixture can be injected directly.

Instrumental Conditions

A wide range of column conditions have been used in the separation of sugar derivatives [37], and the conditions described below are those found to give good resolution.

Trimethylsilyl Derivatives

Using an F & M temperature-programmed apparatus.

Column packing. 3 % OV1 on gas chrom Q; 4 ft column.
Column temperature. 165 ∼ 250 °C (rate of change 3 deg C/min).
Carrier gas flow. 40 ∼ 60 ml/min.

Notes

Separation of lactose anomers from sucrose is difficult and the best results are obtained with two runs, one before and one after inversion of sucrose with invertase.

Alditol Acetates

Column packing. 3 % ECNSS-M on gas chrom Q; 4 ft column.
Column temperature. 195 °C isothermal.
Carrier gas flow. 60 ml/min.

Notes

This gives excellent resolution.

STARCH BY DIRECT ACID HYDROLYSIS [103]

This procedure can be used when only starch (and dextrins) is known to be present, *i.e.* no other easily hydrolysable polysaccharides are present, or where a lower level of accuracy is permissible.

Method. The sample should be finely ground or dispersed in such a way that aliquots of the mixture can be taken accurately. Free sugars should ideally have been removed before this procedure, which is conveniently carried out on the residue insoluble in 80 % (v/v) ethanol.

Measure portions of the sample containing between 100 and 400 mg starch into 1 litre round-bottomed flasks and add 450 ml of about 0·36N H_2SO_4 (1 %, v/v), then boil the mixture under reflux for 4 h. After 30 min, swirl the contents of the flask to rinse any particles of sample from the walls of the flask back into the hydrolysing acid. This procedure may need to be repeated. At the conclusion of the hydrolysis, allow the mixture to cool and then partially neutralise it with sodium hydroxide solution (about 15 ml 10N) and make to 500 ml. Filter the mixture and completely neutralise it with solid sodium carbonate before analysis.

SELECTIVE EXTRACTION OF STARCH

Calcium Chloride Solution for Polarimetry
This forms the basis of a well-established procedure for the measurement of starch in cereal foods, especially flours [43].

Reagents

Acid calcium chloride. Dissolve 620 g of $CaCl_2 . 6H_2O$ in 180 ml water and filter until clear. Add a solution containing 18 g of sodium acetate trihydrate in 50 ml water to the clear filtrate and adjust the pH of the mixture to 2·3 by the addition of acetic acid. Adjust the specific gravity of the solution to 1·30 at 20 °C. (If a deposit forms in cold weather it should be redissolved by warming.)

Carrez's solutions. Solution I: Dissolve 21·9 g zinc acetate dihydrate and 3·0 ml acetic acid in 100 ml water. Solution II: Dissolve 10·6 g potassium ferrocyanide in 100 ml water.

Method

Mix 2·50 g of sample to a smooth paste with 10 ml of water in a tall 400 ml beaker. Add 50 ml calcium chloride solution and autoclave at 103·42 kN/m² for 10 min.

Cool the mixture by immersion in cold water and transfer to a 100 ml volumetric flask using additional calcium chloride solution until the volume is approximately 90 ml. Add 2 ml of Carrez's solution I and shake the mixture well before adding 2 ml of Carrez's solution II. Shake the mixture again and then adjust to 100 ml with calcium chloride solution.

Filter through Whatman no. 541 filter paper; the filtrate should be clear.

Discard the first 15–20 ml of filtrate and take polarimeter readings on the subsequent filtrate.

Calculation. If the reading at 20 °C in a 2 dm tube is $n°$ and the $[\alpha]_D^{20}$ is 203, then

$$\text{starch (g)}/100\,\text{g} = \frac{n \times 10^4}{203 \times 5}$$

Perchloric Acid and Precipitation of Iodine Complex [94]
The starch is extracted from the dried sample with perchloric acid and is then precipitated as the iodine complex, which is decomposed before the starch is hydrolysed.

Reagents
Perchloric acid. 72 % solution, $11\cdot2 \sim 11\cdot4$N.
Alcoholic NaCl. Mix 350 ml of ethanol with 80 ml of water and 50 ml of 20 % (w/v) NaCl and make up to 500 ml with water.
NaCl. 20 % (w/v).
Iodine in potassium iodide. Dissolve $7\cdot5$ g I_2 and $7\cdot5$ g potassium iodide in 150 ml of water, filter and make up to 250 ml.
Alcoholic NaOH. Mix 350 ml of ethanol and 100 ml of water with 25 ml 5N NaOH and make up to 500 ml with water.

Extraction
Measure a portion of the sample (50 $\sim$ 250 mg) into a test-tube with some fine sand and add 4 ml of water.

Heat the tube for 15 min in a boiling water-bath to gelatinise the starch. Cool the tube to room temperature and add 3 ml of perchloric acid rapidly with constant agitation. Grind the sample with a stout glass rod for a minute or so and at intervals for the next 15–20 min. Rinse the rod with 20 ml of water, mix the contents of the tube and centrifuge. Decant the clear supernatant and re-extract the residue with 4 ml of water and 3 ml of perchloric acid.

Make the combined extracts up to 50 ml and ideally analyse at once.

Precipitation of the complex. Transfer a portion of the perchloric acid extracts (1–10 ml) to a tube calibrated at 10, 15 and 20 ml, and dilute to 10 ml. Add 5 ml of 20 % (w/v) NaCl, followed by 2 ml of iodine in KI. Leave the mixture for 20 min. Centrifuge and discard the supernatant. Resuspend the precipitate in 5 ml of alcoholic NaCl and wash by centrifugation.

Decomposition of the complex. Decompose the complex by the addition of 2 ml of alcoholic NaOH to the precipitate. Centrifuge the liberated starch and wash with 5 ml of alcoholic NaCl.

Measurement. The starch may be hydrolysed and the glucose measured or it may be measured directly with the anthrone reagent.

Extraction with Perchloric Acid and Reaction with Anthrone [26]

The method was originally developed for cereals and involves the extraction of free sugars with aqueous ethanol and extraction of the starch with perchloric acid from the residue. The extracted starch is then measured directly, after dilution, by the anthrone reaction.

Reagents

Ethanol. 80 % (v/v).

Perchloric acid. 52 %, prepare by adding 270 ml of 72 % perchloric acid to 100 ml of water.

Anthrone reagent. Dissolve 1 g anthrone in 1 litre of sulphuric acid containing 760 ml concentrated H_2SO_4. Prepare the reagent fresh each day.

Glucose standards. A single standard equivalent to 100 μg of glucose/ml was used originally. A series of standards 25, 50 and 100 μg/ml would be preferable.

Method

Extraction of sugars. Weigh approximately 0·2 g of finely ground sample into a 50 ml centrifuge tube, add two drops of 80 % (v/v) ethanol to moisten the sample then 5 ml of water and stir thoroughly. Add 25 ml of hot 80 % (v/v) ethanol, stir thoroughly, set aside for 5 min and then centrifuge. Decant the supernatant and repeat the extraction with 30 ml of 80 % (v/v) ethanol. Combine the two extracts for the measurement of sugars after removal of the ethanol by evaporation at reduced pressure.

Extraction of starch. Add 5 ml of water to the residue after alcoholic extraction, followed by 65 ml of 52 % (v/v) perchloric acid while stirring the mixture. Continue stirring for 5 min after the addition of the perchloric acid and at intervals for the next 15 min. Add 20 ml of water and centrifuge the mixture. Decant the supernatant into a 100 ml volumetric flask and re-extract the residue as before. Add the contents of the tube to the flask containing the first extract and make to volume with water. Filter the mixture, discarding the first 5 ml of the filtrate. Dilute a portion of the filtrate to give a glucose concentration of around 100 μg glucose per ml.

Analysis of the extracts. The anthrone method used by Clegg involved taking 1 ml of sample extract plus 1 ml of water in tubes with rubber stoppers and adding 10 ml of anthrone/sulphuric acid. The contents of the tube are mixed and heated for 12 min in a boiling water-bath. The absorbance of the solution is measured at 607 nm.

Notes
This method gives quite a good measure of sugars plus starch in cereals: a value which was highly correlated with the metabolisable energy of cereal mixtures for poultry and hence with available carbohydrates.

Dimethyl sulphoxide [20]
Extraction with this reagent is recommended for preparing extracts for amyloglucosidase hydrolysis.

Reagents
  Dimethyl sulphoxide. Spectroscopic grade.
  HCl. 8N.
  Extraction. The sample should be finely divided. Mix portions of about 100 mg with 20 ml of dimethyl sulphoxide and 5 ml of 8N HCl then heat in a water-bath at about 60 °C for 30 min. Dilute the mixture with 50 ml of water and neutralise with 5N NaOH. Cool to room temperature and make to 100 ml. Filter the solution if it is not clear. Clear or faintly opalescent solutions prepared in this way may be treated directly with amyloglucosidase and the glucose formed measured by any suitable enzymatic procedure.

ENZYMATIC HYDROLYSIS OF STARCH

Several enzyme preparations are available which have been used in the analysis of starch. The earlier preparations of the Takadiastase type contained a mixture of amylases, and some gave only a partial conversion of starch to glucose; others, such as Takadiastase (for analysis on Talc, Parke-Davis), gave almost theoretical conversion of starch to glucose [103].

For routine analytical work it is necessary to have an enzyme preparation that is reasonably consistent from batch to batch and that has a well-defined specificity—the amyloglucosidase preparations from *Aspergillus niger* (from Sigma or Boehringer) would satisfy these criteria. The methods described are based on the use of this enzyme.

Preparation of the Sample
Adequate disruption of the starch granules and dispersion of the starch is essential. This is aided if the sample has been treated with a lipid solvent or hot 85% (v/v) methanol to remove the lipoprotein coat of the starch granule.

Gelatinisation by treating the sample with water in a boiling water-bath for about 10 min is adequate for many foods. Very hard-baked products may require brief treatment in an autoclave to disrupt the granules.

Some methods specify a short preliminary acid treatment but this will hydrolyse the more labile polysaccharides (glucofructans, and arabinosyl side-chains of some non-cellulosic polysaccharides) and can be criticised on this account.

Takadiastase

This procedure is based on the method of Southgate [103] using a Takadiastase preparation, but amyloglucosidase (either Boehringer or Sigma) could be substituted.

Method

Weigh a sample of the air-dry residue insoluble in 80 % (v/v) ethanol or 85 % (v/v) methanol into a conical flask and add 5 ml of water. Heat the mixture in a boiling water-bath for 10–15 min to gelatinise the starch.

Cool the mixture to about 40 °C and add 1·2 ml of 2M acetate buffer (pH 4·5), followed by 5 ml of 5 % (w/v) Takadiastase (Takadiastase for analysis on Talc, Parke-Davis) (1 ml of 1 mg/ml suspension of amyloglucosidase). Add a few drops of toluene and incubate the mixture at 37 °C overnight.

At the conclusion of the incubation add 4 volumes of ethanol to the enzymic mixture and mix the contents of the flask. Centrifuge the mixture and decant the clear supernatant into a 100 ml volumetric flask. Re-suspend the precipitate in 20 ml of 80 % (v/v) ethanol and recentrifuge. Make the combined supernatants to volume and measure the glucose formed. In the original procedure either the alcohol was removed and reducing sugars measured or the glucose was measured by the anthrone reaction.

The anthrone procedure has the advantage that the alcohol removal stage is unnecessary and the presence of small amounts of maltose in the enzymatic hydrolysate is of no consequence.

Any procedure for glucose could be used, although the enzymatic methods would also require the removal of alcohol.

Amyloglucosidase

This is a completely enzymatic procedure proposed by Boehringer [20].

Method

Extract 100 mg ground sample with 20 ml of dimethyl sulphoxide and 5 ml of 8N HCl in a water-bath at about 60 °C for 30 min (*see* p. 127).

Dilute the clear supernatant to give around 0·2 ~ 0·4 g starch/litre. Mix

0·2 ml of this with 0·20 ml of acetate buffer (0·1 M, pH 4·6). Then add 0·02 ml of amyloglucosidase (10 mg/ml) and incubate the mixture for 15 min at 20–25 °C.

After this time measure the glucose by the hexokinase method (*see* this chapter (Biochemical procedures)).

Notes
The initial extraction with dimethyl sulphoxide and acid will result in degradation of some polysaccharides other than starch. The amyloglucosidase preparation hydrolyses 1, 4 and 1, 6α-glucosidic bonds and α-1, 2 bonds to a certain extent.

FIBRE METHODS

These methods are widely used in food analysis and, while they do not measure food carbohydrates in any strict sense, they are widely used to obtain estimates of the indigestible matter in foods derived from the structural components of the plant cell wall.

The three selected methods described are, first, the *Crude Fibre* (CF) procedure in its classical form. The use of this method is not recommended, as will be evident from the earlier discussion, nevertheless its wide use means that its omission is undesirable. The other two procedures, for *Neutral Detergent Fibre* (NDF) and *Acid Detergent Fibre* (ADF) are less empirical, and in their latest forms provide the basis for the measurement of cellulose.

Crude Fibre
This method is described in detail because it is still widely used, particularly as a statutory method. As such it must therefore be regarded as a procedure that the food analyst should understand and, if the need arises, be able to execute. However, it is an unsatisfactory method from many points of view and, unless a value for 'fibre' by an official standard procedure is required, the analyst is advised to use the acid-detergent fibre method (*see* below).

The method is very empirical and strict adherence to the specified analytical protocol is essential if repeatable results are to be obtained. The method described is that of the AOAC for 'Crude Fiber' [4].

In this 'crude fibre is the loss of ignition of the dried residue remaining after digestion of the sample with 1·25% H_2SO_4 and 1·25% NaOH solutions under specified conditions'. The method is applicable to grains, meals, flours, animal feeds and all materials containing fibre from which any fat can be extracted to give a workable residue.

Reagents

Sulphuric acid. 0·255N; 1·25 g H_2SO_4/100 ml.

Sodium hydroxide. 0·313N; 1·25 g NaOH/100 ml. This solution should be free or virtually so from carbonate. The strengths of these solutions must be checked by titration and the alkali must be stored in a bottle protected from atmospheric carbon dioxide.

Asbestos. The asbestos must be treated in such a way that the loss on ignition and with the two reagents is reduced to a very low level. The AOAC recommend ignition at 600 °C in a furnace, followed by 30 min in boiling 1·25 % H_2SO_4, filtration and washing with water. The asbestos is then boiled for 30 min in 1·25 % NaOH and filtered and washed with acid, then with water. It is then dried and re-ignited at 600 °C for 2 h.

A blank on 1 g of this asbestos should be carried through the entire method and the blank value (around 1 mg) should be used to correct the crude fibre values obtained with that batch of asbestos.

Alcohol. Methanol, isopropanol or 95 % (v/v) ethanol can be used.

Anti-foaming agent. An anti-foam A (Dow Canning Corporation) diluted 1 + 4 with petroleum spirit is suitable.

Boiling stone. Anti-bumping granules that do not give any blank with the method.

Special Apparatus

Digestion apparatus. For the most efficient operation of this method, a digestion apparatus with condensers to fit 600 ml beakers and with a hot-plate that will bring 200 ml water at 25 °C to a rolling boil in 15 ± 2 min is desirable.

Filtering apparatus. A stainless steel screen (200 mesh) is ideal, as it can be easily cleaned after use.

This can be used as in the Oklahoma State filter, which is lowered into the beaker and the filtrate is withdrawn vertically. Alternatively it can be mounted in a suitable Büchner funnel. In either case the suction flask must be connected to a vacuum line fitted with a valve for breaking the suction.

Reagent preheater. This is needed to maintain the acid and alkali at the temperature of boiling water.

Preparation of Sample

The sample must be finely ground to a uniform consistency: extract 2 g of the ground material with diethyl ether or petroleum spirit to remove fat. (This stage can be omitted if the fat content is less than 1 %.)

Method
Transfer the fat-free material to a 600 ml beaker; add 1 g of the prepared asbestos, 200 ml of boiling 1·25% H_2SO_4 and one drop of anti-foaming agent together with anti-bumping granules. Place the beaker on a preheated hot-plate and boil for exactly 30 min, rotating the beaker at intervals to keep solids from adhering to the sides of the beaker.

Filter the contents of the beaker. Wash the beaker at once with 50–75 ml of boiling water without breaking the suction. When this wash has been removed continue the washing with three further 50 ml portions of water. The filtration should be carried out rapidly and the filter mat should not be allowed to dry until the last washing is removed.

Return the filter mat and residue to the beaker and add 200 ml of boiling 1·25% NaOH. Boil the mixture for exactly 30 min and then filter as before. Wash the filter mat once with 25 ml of boiling 1·25% H_2SO_4, then three times with 50 ml portions of water and 25 ml of alcohol and suck dry.

Transfer the mat and residue to an ashing dish and dry for 2 h at 130 °C ± 2 deg C, cool in a desiccator and weigh.

Heat the dish and contents for 30 min at 600 °C ± 15 deg C. Cool in a desiccator and reweigh.

Calculation. Crude fibre is the loss of ignition minus the loss of the asbestos blank.

Comments
This method is technically exacting and considerable practice is usually required of an analyst who hopes to obtain consistent and reliable results.

Strict attention to detail is essential, particularly the times of heating and the filtration procedures and these especially require practice in order to complete them correctly.

The results obtained are only approximate measures of the cellulose and lignin in a foodstuff and, strictly speaking, the method measures the empirical fraction of a food defined by the method itself.

Detergent Fibre Methods
Two methods are described; the first, for Neutral Detergent Fibre (NDF), is based on extraction of the food with the hot solution of a neutral detergent and the second, Acid Detergent Fibre (ADF), on the extraction with hot, acid-containing detergent. The descriptions are based on the publication by Goering and Van Soest [49], which in turn is based on the work of Van Soest [119, 120].

Neutral detergent fibre is thought by Van Soest to give a good measure of the total cell wall material in a forage and to measure the components of the plant cell wall that require the intervention of micro-organisms for their digestion. Acid Detergent Fibre, on the other hand, gives a good measure of the cellulose and lignin in a food and can be used for the subsequent measurement of these two components [122].

Apparatus

Refluxing apparatus. An apparatus with which at least 6 digestions under reflux can be carried out simultaneously is desirable. Each digestion needs an individual hot-plate or burner that can be regulated and has sufficient power to maintain the boiling solutions at a gentle rolling boil. The reflux condensers should be straight-sided and capable of maintaining the volume of the boiling solutions.

Filtration apparatus. Goering and Van Soest [49] describe a filtration manifold on which several filtrations can be carried out simultaneously. Six Büchner funnels are connected by heavy-walled rubber tubing to a large polythene pipe (60 cm long, 2·5 cm internal diameter), on which six polythene tubes have been welded. The large pipe is connected to the vacuum supply and the level of suction can be adjusted by a valve at the other end of the pipe (Fig. 8.1).

Filter crucibles. Gooch-type Pyrex glass crucibles (50 ml) fitted with coarse sintered filters. These fit into rubber adaptors in the Büchner funnels on the manifold.

Preheater for water. A system for maintaining a supply of hot water for washing the crucibles is desirable for maximum efficiency.

Neutral Detergent Fibre

Reagents

Neutral Detergent Reagent
  Disodium ethylene diamine tetra acetate dihydrate (EDTA). 18·61 g.
  Sodium borate decahydrate. 6·81 g.
  Sodium lauryl sulphate (SLS). 30 g.
  2-Ethoxy ethanol. 10 ml.
  Disodium hydrogen phosphate. 4·56 g.
  Place the EDTA and borate in a large beaker and add some water. Heat the mixture until the reagents dissolve, then add the SLS and 2-ethoxy ethanol.
  Dissolve the phosphate in water in another beaker by heating and then add the solution to the other reagents. Make up to 1 litre. The pH should be

between 6·9 and 7·1 but the need for adjustment is unusual if the solution has been correctly prepared.

Decahydronaphthalene. Reagent grade.

Acetone. Any grade is suitable if colourless and leaving no residue on evaporation.

Sodium sulphite. Anhydrous (reagent grade).

Asbestos. The asbestos should be thoroughly washed with water and acid and ignited at 800 °C before use.

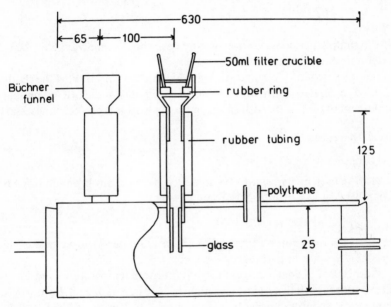

Fig. 8.1.   Sketch of filter manifold construction. Dimensions are shown in mm.
Only 3 of the 6 outlets are shown.

Method

Weigh 0·5 ~ 1·0 g air-dry sample (ground to pass a 1 mm mesh) into a beaker that fits onto the digestion apparatus (normally 600 ml tall form without spout).

Add 100 ml of neutral detergent solution (at room temperature) followed by 2 ml of decahydronaphthalene and 0·5 g sodium sulphite. Heat to boiling in 5 ~ 10 min. Reduce the heating as boiling commences in order to avoid foaming. Adjust the heating to give an even level of boiling and reflux for 60 min, timed from the onset of boiling.

Filter through a previously weighed Gooch crucible swirling the beaker to suspend the solids. Do not apply vacuum until the crucible is full. Filter with a low vacuum at first and increase it only when necessary. Rinse the residue in the beaker with the minimum of hot water (90 ~ 100 °C). Remove the vacuum, stir the residue and fill the crucible with hot water and re-filter.

Wash the crucible and contents with acetone and suck dry. Dry the crucible at 100 °C for 8 h. Cool and weigh.

The ash in the residue can be measured after ignition at 500–550 °C for 3 h.

Notes

If the sample is rich in fat it is usually desirable to remove the fat before attempting to measure NDF.

Samples rich in starch may also give high values for NDF. If the residue gives a strong reaction with iodine in potassium iodide, a preliminary treatment of the sample with an enzyme to hydrolyse the starch is necessary.

Acid Detergent Fibre

Reagents

Acid detergent reagent. Cetyl trimethyl ammonium bromide (CTAB). Dissolve 20 g in 1 litre of $N H_2SO_4$.

Method

Weigh 1 g air-dry sample (ground to pass a 1 mm mesh) into a beaker that fits the digestion apparatus.

Add 100 ml of acid detergent reagent (at room temperature) and 2 ml of decahydronaphthalene. Heat to boiling in 5–10 min, reducing the heating as boiling commences in order to avoid foaming. Boil under reflux for 60 min from the onset of boiling, adjusting the heat to give a slow even boil.

Filter on a previously weighed Gooch crucible using light suction. Wash the crucible and filter mat with hot water (90 ~ 100 °C) after breaking up the filter mat with a rod.

Finally wash with acetone in the same way until it removes no more colour, breaking up any lumps to ensure a complete wash.

Suck the filter dry and dry the crucible for 8 h at 100 °C or overnight, cool and weigh.

Notes

ADF includes all the lignin and cellulose in the sample, together with some inorganic matter. The silica in some ADF residues, notably those from grasses, can be quite high and the ash content of these ADF residues should be measured.

The residue obtained by this method can be used for the measurement of lignin and cellulose and, if the residues are to be used in this way, it is often useful to give a final wash with hexane before drying.

METHOD FOR THE ANALYSIS OF THE CELL WALL COMPONENTS OF ROUGHAGES [45]

This method is described as an example of a detailed analysis of cell wall polysaccharides. The principles followed for the fractionation have been reviewed earlier (p. 70).

Reagents
  Ethanol. 90 % (v/v).
  Ethanol/benzene. 1:2 (v/v).
  Pepsin. 0·5 % (w/v) in 0·1N HCl.
  Ammonium oxalate. 0·5 % (w/v) aqueous solution.
  Potassium hydroxide. 5 % (w/v) and 24 % (w/v).
  Sulphuric acid. N, 4N; 3·5 and 72 % (w/v).

Method
The sample is dried at 100–105 °C in an oven with an inert stream of gas passing through it.

Soluble Carbohydrates
  Mono- and disaccharides. Boil 1 g of the air-dry, powdered sample under reflux in 90 % (v/v) ethanol for 1 h. Filter off the residue and repeat the extraction. Combine the filtrates and measure free sugars (in the original procedure by paper chromatography with rhamnose as an internal standard).

  Fructans. Shake the residue insoluble in ethanol with cold water for 6 h, filter off the insoluble material and repeat the extraction.

  Adjust the combined filtrates to $0·1N H_2SO_4$ by the addition of $4N H_2SO_4$ and heat on a boiling water-bath for 1 h. Neutralise the solution with solid barium carbonate, concentrate under reduced pressure and measure the glucose and fructose.

  Structural carbohydrates. Extract the original dried material with ethanol/benzene in a Soxhlet apparatus until the lipids and pigments have been removed. Dry and weigh the residue.

Lignin. Digest a 1 g portion overnight with 40 ml of pepsin in HCl solution at 40 °C. Add hot water (30 ml) and filter the mixture through a glass filter with a layer of asbestos on the sintered filter. Wash the residue twice with hot water and then boil under reflux with 150 ml of 5 % (w/v) H_2SO_4 for 2 h. Filter the mixture and wash the residue three times with 30 ml of hot water and dry with ethanol and diethyl ether.

Add 20 ml of 72 % (w/v) H_2SO_4 to the residue, and maintain the mixture at 18 °C $\pm$ 2 deg C for 2 h. After this time add 125 ml of water, filter the mixture and wash the residue with water. Boil the residue under reflux in 3 % (w/v) H_2SO_4 for 2 h, filter on a weighed ash-less filter paper. Dry and weigh the paper and residue. Ignite the paper and weigh the ash. The loss in weight is taken as lignin.

Pectic substances. Weigh 3 g of ethanol/benzene-extracted material into a glass filter crucible and extract with 0·5 % (w/v) ammonium oxalate for 2 h at 85 °C. Repeat the extraction four times. Combine the filtrates, make them slightly acid with N hydrochloric acid and add 4 volumes of ethanol with stirring. After the precipitate has settled, decant the supernatant through a weighed glass filter crucible. Transfer the precipitate to the crucible and wash with 70 % (v/v) ethanol (slightly acidified with HCl), then with ethanol followed by acetone. Suck the crucible dry, dry completely at 100 °C, cool and weigh.

In the original procedure, if the precipitate weighed less than 100 mg its weight was taken as pectic substances (< 3 % of ethanol/benzene-insoluble residue); if the amount recovered was greater than this it was hydrolysed under reflux with N H_2SO_4 and the sugars in the mixture measured by paper chromatography. Uronic acids were measured by decarboxylation with 12 % (v/v) HCl and absorption of the carbon dioxide in barium hydroxide.

Hemicelluloses. If the sample contains less than 6 % lignin, the hemicelluloses may be extracted directly. Place the residue after extraction of pectic substances in a 100 ml conical flask and add sufficient 5 % (w/v) KOH to cover the residue. Then flush the flask with nitrogen and shake the mixture for 24 h. Filter off the residue on a sintered glass crucible and repeat the extraction three times. Make the combined filtrates faintly acid with acetic acid and mix in 4 volumes of ethanol. Collect the precipitate in a glass filter crucible, wash with acetone, dry and weigh.

Extract the residue insoluble in 5 % KOH in the same way with 24 % KOH. Hydrolyse portions (50 mg) of the residue with N H_2SO_4 and measure sugars in the hydrolysates.

If the material is straw, delignify as indicated below, but with other materials containing more than 6 % lignin, first extract with 5 % KOH and then delignify before the second extraction with 5 % KOH.

Delignification. Delignify the residue after extraction with ammonium oxalate by the addition of chloramine and acetic acid. Treat the residue in 75 ml of water in a boiling water-bath with about 0·5 ml acetic acid and 1 g chloramine for 2 h. Filter the mixture through a sintered glass filter and wash twice with ethanol. Extract the residue twice with a boiling 3 % (v/v) solution of ethanolamine in ethanol and wash the residual holocellulose with ethanol; finally extract the holocellulose with alkali as before.

Cellulose. Take the residue insoluble in 24 % (w/v) KOH and wash thoroughly with water followed by acetone, dry and weigh. Hydrolyse in $N H_2SO_4$ and filter the residue on a weighed ash-less filter paper, wash the paper with water, acetone, and then dry and reweigh. Finally ash and weigh the residue. Deduct the weight of ash from the weight of the residue after dilute acid hydrolysis and take the value obtained as cellulose.

Comments
This method is far too time-consuming to be used as a routine analytical procedure; however, it does provide values for the complete analysis of the structural carbohydrates and lignin in plant tissues.

THE MEASUREMENT OF UNAVAILABLE CARBOHYDRATES OR DIETARY FIBRE IN FOODS

The procedures described in these working instructions are modifications of those originally published in 1969 [104]. These modifications have been developed in the Dunn Nutritional Laboratory as a result of experience with the methods in semi-routine use. In general, these modifications result in some saving of time over the original procedures.

As a general cautionary note, it is important for the intending analyst to bear in mind that the plant cell wall is composed of a mixture of complex polysaccharides and many other constituents. The initial analyses of a foodstuff, therefore, need to be accompanied by a number of quantitative checks, which it is possible to omit as experience with the foodstuff is accumulated. The analyst must use his intelligence and knowledge in using the methods if errors of interpretation are to be avoided.

These instructions are divided into the following sections, corresponding to the stages in the analysis.

(1) Preparation of the sample.
(2) Extraction of free sugars and preparation of the residue insoluble in 85% v/v methanol.
(3) Enzymatic hydrolysis of starch.
(4) Extraction of water-soluble material.
(5) Hydrolysis with dilute acid.
(6) Extraction of cellulose.

Preparation of Sample
The sample for analysis must be representative and include all the edible material. Heating can often lead to the formation of condensation products between phenolic constituents and protein, which analyse as lignin. When it is necessary to dry a food, freeze-drying is the method of choice. Most fresh foods can be extracted directly with 85% (v/v) methanol.

Homogenisation of the sample should be carried out after freeze-drying to avoid enzymatic modifications that may occur when the cells are disrupted.

Extraction of Free Sugars and Preparation of a Residue Insoluble in 85% (v/v) Methanol
The food, either fresh or after freeze-drying, is extracted with hot 85% (v/v) methanol followed by hot diethyl ether. This removes free sugars, lipids, most pigments and cuticular waxes, etc.

Reagent. 85% (v/v) aqueous methanol.

Procedure
Weigh out duplicate portions of between 3 and 5 g (dry weight or equivalent) into weighed 100 ml beakers and add 25 ml 85% (v/v) methanol[1] (*see* Notes below). Stir with a glass rod while bringing the mixture to the boil on an electric hot-plate.[2] Filter hot through a Whatman 541 paper (if sugars are to be measured in the sample the filtration is made into a 100 ml volumetric flask). Repeat this process with 3 further portions of aqueous methanol, allowing the filter to drain between successive extractions. In the second and subsequent extractions it is *essential* to stir the mixture *continuously* while it is being heated to avoid losses through 'bumping'.

Then extract the residue in the beaker and on the filter in a similar way with three portions of diethyl ether, discard the extracts.

Allow the residue to dry in air and return to the original beaker.[3] Heat the beaker and contents at 98–100 °C for 10 min in an air oven to remove residual solvent, allow to cool in air overnight[4] and then reweigh. The residue should then be finely ground and stored in an air-tight container.

Notes

(1) If a fresh sample is being extracted, adjust the concentration of methanol added initially so that the first extraction is made at 85 % (v/v).

(2) This can be carried out in a fume cupboard but in practice the amounts of solvent released are quite small. When using diethyl ether it is essential to ensure that spark-free apparatus is used.

(3) The residue should only be returned to the beaker when it is dry and crumbles easily. Scrape the filter paper gently with a spatula to ensure quantitative transfer.

(4) This ensures that the residue is in equilibrium with the laboratory atmosphere, and can be manipulated without fear of its moisture content changing. Under normal conditions, residues prepared in this way contain about 8–10 % moisture.

Calculation. The residue insoluble in 85 % (v/v) methanol is expressed as a proportion of the original sample.

Fractionation of Polysaccharides in the Residue Insoluble in 85 % (v/v) Methanol

Enzymatic Hydrolysis of Starch

Reagents

2M *acetate buffer.* pH 4·6.

Takadiastase (for analysis on Talc). Shake 10 g with 100 ml water, centrifuge and use supernatant.

Procedure

Weigh out duplicate portions of about 200–300 mg (ideally containing approximately 100 mg starch) into a clean, dry 50 ml centrifuge tube. Add 4 ml of hot water and stir with a glass rod. Place the tubes in a boiling water-bath for 10 min to gelatinise the starch in the sample. Cool and then add 0·2 ml of 2M acetate buffer (pH 4·6) followed by 1 ml of 10 % Takadiastase. Mix the contents of the tube with the rod and 'police down' any particles above the level of the liquid. Add a few drops of toluene and incubate at 37 °C overnight.

The next morning add 4 volumes of ethanol (20 ml), performing the addition in two stages, stirring the mixture thoroughly after the first addition and finally rinsing the rod with the last of the ethanol. Leave the mixture for about 10 min to allow the precipitate to form then centrifuge for 10 min (at 2500 rev/min). Pour off the supernatant carefully into a volumetric flask

(100 ml if the sample contains appreciable amounts of starch, 50 ml if very low). Wash the residue by resuspending it in 10 ml of 80 % (v/v) ethanol and recentrifuge. Pour off the supernatant and combine it with the previous supernatant.

Make up the combined supernatants to volume and label these as S (starch) extracts. Glucose is measured in these extracts.

Extraction of Water-soluble Material
Heat the tubes containing the residue after treatment with Takadiastase in a boiling water-bath to drive off the ethanol. (Addition of a few ml of water at this stage is useful.) When the tubes are free of ethanol, add 10 ml of hot (near boiling) water and stir with a glass rod. Leave the tubes in the water-bath for 20 min, then centrifuge and pour off the supernatant. Repeat the extraction and combine the supernatants.

When the supernatants are cool, add 4 volumes of ethanol and centrifuge. Wash the precipitate by resuspension in 80 % (v/v) ethanol and centrifuge.

Resuspend the residue in 10 ml of N H_2SO_4 and heat in a boiling water-bath for $2\frac{1}{2}$ h. Dilute the hydrolysates to volume and measure monosaccharides and uronic acids. Label the extracts W (water) extract.

N.B. Benzoic acid is added to the extracts as a preservative at this and all subsequent stages.

Hydrolysis with Dilute Acid
Suspend the residue after water extraction in the centrifuge tube with 10 ml of N H_2SO_4; cover the tube with a marble or perforated polythene and heat the tube in a boiling water-bath for $2\frac{1}{2}$ h. Mix the contents of the tube after 1 h. (If particles of the residue have been carried up the walls of the tube it may be necessary to rinse them back with additional acid.)

Allow the tubes and contents to cool, add an equal volume of ethanol and then centrifuge or filter the contents. (See Note 1, p. 141.)

Centrifugation
Pour the supernatant carefully into a volumetric flask and wash the residue by resuspension in 50 % (v/v) ethanol and recentrifuge. Repeat this washing procedure and combine the supernatants. Finally wash the residue in the same way with ethanol and diethyl ether and allow the residue to dry in air.

Filtration
Filter the hydrolysate through a weighed sintered glass filter crucible (porosity 1^2). Wash the residue in the filter thoroughly with 50 % (v/v)

ethanol and transfer the aqueous alcoholic filtrate to a volumetric flask. Finally wash the residue with ethanol and diethyl ether and suck the crucible dry. Heat the crucible for 10 min at 98–100 °C, cool and weigh.

Dilute the aqueous supernatant or filtrate to volume and measure hexoses, pentoses and uronic acids.

Notes
(1) Some materials are extremely difficult to filter at this stage; centrifugation is the only practical procedure for these. It is more time consuming.

(2) Many foods yield some very fine particles at this stage and a lower porosity may be desirable, this will slow the process considerably. The analyst will usually have to assess the magnitude of the error produced by the fine particles passing through the filter in relation to the additional time involved in using a finer filter.

Extraction of Cellulose
Reagents. 72 % (w/w) H_2SO_4.

Procedure
Add chilled 72 % (w/w) H_2SO_4 to the dry residue (either in a centrifuge tube or on the sintered glass crucible) after dilute acid hydrolysis and stir gently. Stand the sintered crucibles in small beakers during this procedure. Leave the residue in contact with the acid for at least 24 h at 0–4 °C, stirring the mixture at intervals during the first few hours.

At the conclusion of this period, either filter the mixture through a sintered glass crucible as described above (*see* Filtration) or rinse the exterior of the crucible into the beaker and wash the residue in the filter very thoroughly with water. Dilute the aqueous filtrates to volume and measure hexoses, pentoses and uronic acids.

Wash the residue on the filter with ethanol followed by diethyl ether. Heat the crucible in an oven and reweigh. The gain in weight is taken as lignin.

DETERMINATION OF PECTIC ACID IN FRUITS AND FRUIT PRODUCTS

In this AOAC procedure [4] the pectic substances are saponified to remove ester groups and the pectic acid is precipitated and weighed.

Reagents
  H_2SO_4. Normal.
  NaOH. 10 % (w/v).
  Ethanol. 95 % (v/v).
  HCl. 10 ml + 25 ml water.

Method
Prepare a hot water extract by extracting 300 g with water on a steam bath. Whole fruits should be chopped thoroughly before extraction. Extract the chopped sample (300 g) with 800 ml of water at the temperature of boiling water for 1 h. Transfer the mixture to a 2 litre flask and make to volume.

Place a 200 ml filtered portion of the extract in a beaker (600 ml) and add 8–12 g sucrose if the solution is low in sucrose. Add 200 ml of alcohol with constant stirring and allow the precipitate to settle. Filter off the precipitate on a qualitative paper and wash with alcohol.

Return the precipitate to its original beaker with hot water and evaporate to a volume of about 40 ml, then cool the solution to 25 °C. Redissolve any precipitate that forms at this stage by acidifying the solution and warming. Dilute 2–5 ml of 10 % (w/v) NaOH to 50 ml and add to the solution. (The volume of NaOH used depends on the bulk of the precipitate.)

Leave the mixture for 15 min, then add 40 ml of water followed by 10 ml of diluted HCl and boil the mixture for 5 min. Filter the precipitated pectic acids and wash with hot water.

Return the precipitate to the beaker and repeat the saponification. Wash the precipitate until the filtrate is only slightly acid. Transfer the precipitate to a platinum dish and dry at 100 °C, weigh and then ignite and reweigh. Take the loss in weight as pectic acid.

Notes
A colloidal precipitate that is difficult to filter is indicative that saponification is incomplete and that it should be repeated.

DETERMINATION OF LIGNIN BY PERMANGANATE OXIDATION [49, 122]

This is an indirect method that can be used to measure lignin in acid detergent fibre (ADF) and so provides a method for deriving cellulose values from ADF. It is recommended as an alternative to the measurement of lignin as the residue insoluble in 72 % (w/w) H_2SO_4.

The advantages of the oxidation procedure are that it is somewhat shorter and uses less corrosive reagents, which do not need standardisation.

More importantly, the method does not measure lignin artefacts produced in foods by heating and may therefore give values for lignin that are closer to 'true lignin'. Cutin is, however, not oxidised by the reagent and a combination of the two procedures is suggested in order to obtain an estimate of acid detergent cutin.

Reagents

Saturated potassium permanganate. Dissolve 50 g $KMnO_4$ (AnalaR) and 0·05 g Ag_2SO_4 in water and make up to 1 litre with water. Store the solution in the dark.

Lignin buffer solution. Dissolve 6 g $Fe(NO_3)_3 . 9H_2O$ and 0·15 $AgNO_3$ in 100 ml water and combine with 500 ml of acetic acid and 5 g potassium acetate. Add 400 ml of *tert* butyl alcohol and mix. All the reagents used should pass the dichromate tests for halogens.

Combined reagent. Mix the $KMnO_4$ solution and lignin buffer in the proportions 2 to 1 (v/v). The unused reagent can be stored for about one week at $0 \sim 4\,^{\circ}C$ in the dark. The solution can be used while it is still purple and has not deposited a precipitate, the formation of a reddish colour is indicative of deterioration.

Demineralising solution. Dissolve 50 g oxalic acid dihydrate in 700 ml of 95 % (v/v) ethanol, add 50 ml of conc. HCl and dilute to 1 litre with water.

Ethanol. 80 % (v/v).

Method

Dry samples at a temperature below 65 °C and grind to pass a 1 mm sieve. Measure ADF by the method described earlier in this chapter, using a 1 g sample (if the food contains more than 15 % lignin the sample size should be reduced to 0·5 g).

Place the weighed crucible, containing the ADF in a shallow enamel dish containing water to a depth of about 1 cm.

Add 25 ml of combined reagent to the crucible and adjust the level of water in the dish to regulate the flow of reagent through the crucible. Place a short glass rod in each crucible to stir the contents, breaking up any lumps and ensuring that all the ADF is thoroughly wetted with the reagent.

Leave the crucibles to stand at $20 \sim 25\ C$ for 90 ± 10 min, adding more combined reagent as required; a purple colour must be maintained in the crucible.

Remove the crucibles and filter. Suck the crucibles dry but do *not* wash the residue.

Add the demineralising solution to the crucible and leave for 5 min, suck dry and then half re-fill the crucible with the solution. Repeat after another 5 min. Rinse the crucible with a thin stream of water. Continue the treatment until the fibre is white, a process which will take between 20 and 30 min.

Fill the crucible with 80 % (v/v) ethanol and suck dry; repeat the washing twice and then wash the residue with acetone in the same way.

Finally dry the crucible at 100 °C overnight and re-weigh.

Lignin is taken as the loss in weight.

Notes

Ash may be determined on the residue and thus provide an estimate of the cellulose in the ADF. The reliability of the values obtained in this way depend on the type of sample. Where the concentration of lignin is low the oxidation is very rapid and losses of cellulose can occur if the oxidation stage is too vigorous. When samples of this kind are being analysed, the rate of flow of the combined reagent through the crucibles should be reduced by adjusting the level of water in the dish.

APPENDIX

Reference Tables

TABLE A.1
SPECIFIC GRAVITY, DEGREES BRIX AND DEGREES BAUMÉ OF SUCROSE SOLUTIONS[a]

Degrees Brix (% sucrose, w/w)	Spec. grav. 20/20°C	Degrees Baumé (modulus 145)	Degrees Brix (% sucrose, w/w)	Spec. grav. 20/20°C	Degrees Baumé (modulus 145)	Degrees Brix (% sucrose, w/w)	Spec. grav. 20/20°C	Degrees Baumé (modulus 145)	Degrees Brix (% sucrose, w/w)	Spec. grav. 20/20°C	Degrees Baumé (modulus 145)
0·0	1·00000	0·00	10·0	1·03998	5·57	20·0	1·08287	11·10	30·0	1·12898	16·57
0·2	1·00078	0·11	10·2	1·04081	5·68	20·2	1·08376	11·21	30·2	1·12993	16·67
0·4	1·00155	0·22	10·4	1·04164	5·80	20·4	1·08465	11·32	30·4	1·13089	16·78
0·6	1·00233	0·34	10·6	1·04247	5·91	20·6	1·08554	11·43	30·6	1·13185	16·89
0·8	1·00311	0·45	10·8	1·04330	6·02	20·8	1·08644	11·54	30·8	1·13281	17·00
1·0	1·00389	0·56	11·0	1·04413	6·13	21·0	1·08733	11·65	31·0	1·13378	17·11
1·2	1·00467	0·67	11·2	1·04497	6·24	21·2	1·08823	11·76	31·2	1·13474	17·22
1·4	1·00545	0·79	11·4	1·04580	6·35	21·4	1·08913	11·87	31·4	1·13570	17·33
1·6	1·00623	0·90	11·6	1·04664	6·46	21·6	1·09003	11·98	31·6	1·13667	17·43
1·8	1·00701	1·01	11·8	1·04747	6·57	21·8	1·09093	12·09	31·8	1·13764	17·54
2·0	1·00779	1·12	12·0	1·04831	6·68	22·0	1·09183	12·20	32·0	1·13861	17·65
2·2	1·00858	1·23	12·2	1·04915	6·79	22·2	1·09273	12·31	32·2	1·13958	17·76
2·4	1·00936	1·34	12·4	1·04999	6·90	22·4	1·09364	12·42	32·4	1·14055	17·87
2·6	1·01015	1·46	12·6	1·05084	7·02	22·6	1·09454	12·52	32·6	1·14152	17·98
2·8	1·01093	1·57	12·8	1·05168	7·13	22·8	1·09545	12·63	32·8	1·14250	18·08
3·0	1·01172	1·68	13·0	1·05252	7·24	23·0	1·09636	12·74	33·0	1·14347	18·19
3·2	1·01251	1·79	13·2	1·05337	7·35	23·2	1·09727	12·85	33·2	1·14445	18·30
3·4	1·01330	1·90	13·4	1·05422	7·46	23·4	1·09818	12·96	33·4	1·14543	18·41
3·6	1·01409	2·02	13·6	1·05506	7·57	23·6	1·09909	13·07	33·6	1·14641	18·52
3·8	1·01488	2·13	13·8	1·05591	7·68	23·8	1·10000	13·18	33·8	1·14739	18·63

4·0	1·01567	2·24	14·0	1·05677	7·79	24·0	1·10092	13·29	34·0	1·14837	18·73
4·2	1·01647	2·35	14·2	1·05762	7·90	24·2	1·10183	13·40	34·2	1·14936	18·84
4·4	1·01726	2·46	14·4	1·05847	8·01	24·4	1·10275	13·51	34·4	1·15034	18·95
4·6	1·01806	2·57	14·6	1·05933	8·12	24·6	1·10367	13·62	34·6	1·15133	19·06
4·8	1·01886	2·68	14·8	1·06018	8·23	24·8	1·10459	13·73	34·8	1·15232	19·17
5·0	1·01965	2·79	15·0	1·06104	8·34	25·0	1·10551	13·84	35·0	1·15331	19·28
5·2	1·02045	2·91	15·2	1·06190	8·45	25·2	1·10643	13·95	35·2	1·15430	19·38
5·4	1·02125	3·02	15·4	1·06276	8·56	25·4	1·10736	14·06	35·4	1·15530	19·49
5·6	1·02206	3·13	15·6	1·06362	8·67	25·6	1·10828	14·17	35·6	1·15629	19·60
5·8	1·02286	3·24	15·8	1·06448	8·78	25·8	1·10921	14·28	35·8	1·15729	19·71
6·0	1·02366	3·35	16·0	1·06534	8·89	26·0	1·11014	14·39	36·0	1·15828	19·81
6·2	1·02447	3·46	16·2	1·06621	9·00	26·2	1·11106	14·49	36·2	1·15928	19·92
6·4	1·02527	3·57	16·4	1·06707	9·11	26·4	1·11200	14·60	36·4	1·16028	20·03
6·6	1·02608	3·69	16·6	1·06794	9·22	26·6	1·11293	14·71	36·6	1·16128	20·14
6·8	1·02689	3·80	16·8	1·06881	9·33	26·8	1·11386	14·82	36·8	1·16228	20·25
7·0	1·02770	3·91	17·0	1·06968	9·45	27·0	1·11480	14·93	37·0	1·16329	20·35
7·2	1·02851	4·02	17·2	1·07055	9·56	27·2	1·11573	15·04	37·2	1·16430	20·46
7·4	1·02932	4·13	17·4	1·07142	9·67	27·4	1·11667	15·15	37·4	1·16530	20·57
7·6	1·03013	4·24	17·6	1·07229	9·78	27·6	1·11761	15·26	37·6	1·16631	20·68
7·8	1·03095	4·35	17·8	1·07317	9·89	27·8	1·11855	15·37	37·8	1·16732	20·78
8·0	1·03176	4·46	18·0	1·07404	10·00	28·0	1·11949	15·48	38·0	1·16833	20·89
8·2	1·03258	4·58	18·2	1·07492	10·11	28·2	1·12043	15·59	38·2	1·16934	21·00
8·4	1·03340	4·69	18·4	1·07580	10·22	28·4	1·12138	15·69	38·4	1·17036	21·11
8·6	1·03422	4·80	18·6	1·07668	10·33	28·6	1·12232	15·80	38·6	1·17138	21·21
8·8	1·03504	4·91	18·8	1·07756	10·44	28·8	1·12327	15·91	38·8	1·17239	21·32
9·0	1·03586	5·02	19·0	1·07844	10·55	29·0	1·12422	16·02	39·0	1·17341	21·43
9·2	1·03668	5·13	19·2	1·07932	10·66	29·2	1·12517	16·13	39·2	1·17443	21·54
9·4	1·03750	5·24	19·4	1·08021	10·77	29·4	1·12612	16·24	39·4	1·17545	21·64
9·6	1·03833	5·35	19·6	1·08110	10·88	29·6	1·12707	16·35	39·6	1·17648	21·75
9·8	1·03915	5·46	19·8	1·08198	10·99	29·8	1·12802	16·46	39·8	1·17750	21·86

^a These figures are accepted by the International Commission for Uniform Methods of Sugar Analysis.

TABLE A.1—*continued*

Degrees Brix (% sucrose, w/w)	Spec. grav. 20/20°C	Degrees Baumé (modulus 145)	Degrees Brix (% sucrose, w/w)	Spec. grav. 20/20°C	Degrees Baumé (modulus 145)	Degrees Brix (% sucrose, w/w)	Spec. grav. 20/20°C	Degrees Baumé (modulus 145)	Degrees Brix (% sucrose, w/w)	Spec. grav. 20/20°C	Degrees Baumé (modulus 145)
40·0	1·178 53	21·97	51·0	1·237 27	27·81	62·0	1·300 59	33·51	73·0	1·368 56	39·05
40·2	1·179 56	22·07	51·2	1·238 38	27·91	62·2	1·301 78	33·61	73·2	1·369 83	39·15
40·4	1·180 58	22·18	51·4	1·239 49	28·02	62·4	1·302 98	33·72	73·4	1·371 11	39·25
40·6	1·181 62	22·29	51·6	1·240 60	28·12	62·6	1·304 18	33·82	73·6	1·372 40	39·35
40·8	1·182 65	22·39	51·8	1·241 72	28·23	62·8	1·305 37	33·92	73·8	1·373 68	39·44
41·0	1·183 68	22·50	52·0	1·242 84	28·33	63·0	1·306 57	34·02	74·0	1·374 96	39·54
41·2	1·184 72	22·61	52·2	1·243 95	28·44	63·2	1·307 78	34·12	74·2	1·376 25	39·64
41·4	1·185 75	22·72	52·4	1·245 07	28·54	63·4	1·308 98	34·23	74·4	1·377 54	39·74
41·6	1·186 79	22·82	52·6	1·246 19	28·65	63·6	1·310 19	34·33	74·6	1·378 83	39·84
41·8	1·187 83	22·93	52·8	1·247 31	28·75	63·8	1·311 39	34·43	74·8	1·380 12	39·94
42·0	1·188 87	23·04	53·0	1·248 44	28·86	64·0	1·312 60	34·53	75·0	1·381 41	40·03
42·2	1·189 92	23·14	53·2	1·249 56	28·96	64·2	1·313 81	34·63	75·2	1·382 70	40·13
42·4	1·190 96	23·25	53·4	1·250 69	29·06	64·4	1·315 02	34·74	75·4	1·384 00	40·23
42·6	1·192 01	23·36	53·6	1·251 82	29·17	64·6	1·316 23	38·84	75·6	1·385 30	40·33
42·8	1·193 05	23·46	53·8	1·252 95	29·27	64·8	1·317 45	34·94	75·8	1·386 60	40·43
43·0	1·194 10	23·57	54·0	1·254 08	29·38	65·0	1·318 66	35·04	76·0	1·387 90	40·53
43·2	1·195 15	23·68	54·2	1·255 21	29·48	65·2	1·319 88	35·14	76·2	1·389 20	40·62
43·4	1·196 20	23·78	54·4	1·256 35	29·59	65·4	1·321 10	35·24	76·4	1·390 50	40·72
43·6	1·197 26	23·89	54·6	1·257 48	29·69	65·6	1·322 32	35·34	76·6	1·391 80	40·82
43·8	1·198 31	24·00	54·8	1·258 62	29·80	65·8	1·323 54	35·45	76·8	1·393 11	40·92
44·0	1·199 36	24·10	55·0	1·259 76	29·90	66·0	1·324 76	35·55	77·0	1·394 42	41·01
44·2	1·200 42	24·21	55·2	1·260 90	30·00	66·2	1·325 99	35·65	77·2	1·395 73	41·11
44·4	1·201 48	24·32	55·4	1·262 04	30·11	66·4	1·327 22	35·75	77·4	1·397 04	41·21

45·0	1·20467	24·63	56·0	1·26548	30·42	67·0	1·33090	36·05	78·0	1·40098	41·50
45·2	1·20573	24·74	56·2	1·26663	30·52	67·2	1·33214	36·15	78·2	1·40230	41·60
45·4	1·20680	24·85	56·4	1·26778	30·63	67·4	1·33337	36·25	78·4	1·40361	41·70
45·6	1·20787	24·95	56·6	1·26893	30·73	67·6	1·33460	36·35	78·6	1·40493	41·79
45·8	1·20894	25·06	56·8	1·27008	30·83	67·8	1·33584	36·45	78·8	1·40625	41·89
46·0	1·21001	25·17	57·0	1·27123	30·94	68·0	1·33708	36·55	79·0	1·40758	41·99
46·2	1·21108	25·27	57·2	1·27239	31·04	68·2	1·33832	36·66	79·2	1·40890	42·08
46·4	1·21215	25·38	57·4	1·27355	31·15	68·4	1·33957	36·76	79·4	1·41023	42·18
46·6	1·21323	25·48	57·6	1·27471	31·25	68·6	1·34081	36·86	79·6	1·41155	42·28
46·8	1·21431	25·59	57·8	1·27587	31·35	68·8	1·34205	36·96	79·8	1·41288	42·37
47·0	1·21538	25·70	58·0	1·27703	31·46	69·0	1·34330	37·06	80·0	1·41421	42·47
47·2	1·21646	25·80	58·2	1·27819	31·56	69·2	1·34455	37·16	80·2	1·41554	42·57
47·4	1·21755	25·91	58·4	1·27936	31·66	69·4	1·34580	37·26	80·4	1·41688	42·66
47·6	1·21863	26·01	58·6	1·28052	31·76	69·6	1·34705	37·36	80·6	1·41821	42·76
47·8	1·21971	26·12	58·8	1·28169	31·87	69·8	1·34830	37·46	80·8	1·41955	42·85
48·0	1·22080	26·23	59·0	1·28286	31·97	70·0	1·34956	37·56	81·0	1·42088	42·95
48·2	1·22189	26·33	59·2	1·28404	32·07	70·2	1·35081	37·66	81·2	1·42222	43·05
48·4	1·22298	26·44	59·4	1·28520	32·18	70·4	1·35207	37·76	81·4	1·42356	43·14
48·6	1·22406	26·54	59·6	1·28638	32·28	70·6	1·35333	37·86	81·6	1·42490	43·24
48·8	1·22516	26·65	59·8	1·28755	32·38	70·8	1·35459	37·96	81·8	1·42625	43·33
49·0	1·22625	26·75	60·0	1·28873	32·49	71·0	1·35585	38·06	82·0	1·42759	43·43
49·2	1·22735	26·86	60·2	1·28991	32·59	71·2	1·35711	38·16	82·2	1·42894	43·53
49·4	1·22844	26·96	60·4	1·29109	32·69	71·4	1·35838	38·26	82·4	1·43029	43·62
49·6	1·22954	27·07	60·6	1·29227	32·79	71·6	1·35964	38·35	82·6	1·43164	43·72
49·8	1·23064	27·18	60·8	1·29346	32·90	71·8	1·36091	38·45	82·8	1·43298	43·81
50·0	1·23174	27·28	61·0	1·29464	33·00	72·0	1·36218	38·55	83·0	1·43434	43·91
50·2	1·23284	27·39	61·2	1·29583	33·10	72·2	1·36346	38·65	83·2	1·43569	44·00
50·4	1·23395	27·49	61·4	1·29701	33·20	72·4	1·36473	38·75	83·4	1·43705	44·10
50·6	1·23506	27·60	61·6	1·29820	33·31	72·6	1·36600	38·85	83·6	1·43841	44·19
50·8	1·23616	27·70	61·8	1·29940	33·41	72·8	1·36728	38·95	83·8	1·43976	44·29

TABLE A.1—*continued*

Degrees Brix (% sucrose, w/w)	Spec. grav. 20/20°C	Degrees Baumé (modulus 145)	Degrees Brix (% sucrose, w/w)	Spec. grav. 20/20°C	Degrees Baumé (modulus 145)	Degrees Brix (% sucrose, w/w)	Spec. grav. 20/20°C	Degrees Baumé (modulus 145)	Degrees Brix (% sucrose, w/w)	Spec. grav. 20/20°C	Degrees Baumé (modulus 145)
84·0	1·441 12	44·38	88·0	1·468 62	46·27	92·0	1·496 71	48·12	96·0	1·525 35	49·94
84·2	1·442 49	44·48	88·2	1·470 02	46·36	92·2	1·498 12	48·21	96·2	1·526 80	50·03
84·4	1·443 85	44·57	88·4	1·471 41	46·45	92·4	1·499 54	48·30	96·4	1·528 24	50·12
84·6	1·445 21	44·67	88·6	1·472 80	46·55	92·6	1·500 97	48·40	96·6	1·529 69	50·21
84·8	1·446 58	44·76	88·8	1·474 20	46·64	92·8	1·502 39	48·49	96·8	1·531 14	50·30
85·0	1·447 94	44·86	89·0	1·475 59	46·73	93·0	1·503 81	48·58	97·0	1·532 60	50·39
85·2	1·449 31	44·95	89·2	1·476 99	46·83	93·2	1·505 24	48·67	97·2	1·534 05	50·48
85·4	1·450 68	45·05	89·4	1·478 39	46·92	93·4	1·506 67	48·76	97·4	1·535 51	50·57
85·6	1·452 05	45·14	89·6	1·479 79	47·01	93·6	1·508 10	48·85	97·6	1·536 96	50·66
85·8	1·453 43	45·24	89·8	1·481 19	47·11	93·8	1·509 52	48·94	97·8	1·538 42	50·75
86·0	1·454 80	45·33	90·0	1·482 59	47·20	94·0	1·510 96	49·03	98·0	1·539 88	50·84
86·2	1·456 18	45·42	90·2	1·484 00	47·29	94·2	1·512 39	49·12	98·2	1·541 34	50·93
86·4	1·457 55	45·52	90·4	1·485 40	47·38	94·4	1·513 82	49·22	98·4	1·542 80	51·02
86·6	1·458 93	45·61	90·6	1·486 81	47·48	94·6	1·515 26	49·31	98·6	1·544 26	51·10
86·8	1·460 31	45·71	90·8	1·488 22	47·57	94·8	1·516 70	49·40	98·8	1·545 73	51·19
87·0	1·461 70	45·80	91·0	1·489 63	47·66	95·0	1·518 14	49·49	99·0	1·547 19	51·28
87·2	1·463 08	45·89	91·2	1·491 04	47·75	95·2	1·519 58	49·58	99·2	1·548 66	51·37
87·4	1·464 46	45·99	91·4	1·492 46	47·84	95·4	1·521 02	49·67	99·4	1·550 13	51·46
87·6	1·465 85	46·08	91·6	1·493 87	47·94	95·6	1·522 46	49·76	99·6	1·551 60	51·55
87·8	1·467 24	46·17	91·8	1·495 29	48·03	95·8	1·523 90	49·85	99·8	1·553 07	51·64
									100·0	1·554 54	51·73

TABLE A.2
FACTORS FOR USE WITH THE LANE–EYNON METHOD
(a) 10 ml Soxhlet reagent

Titre	*Sucrose (g)/100 ml invert sugar*					*Glucose*	*Fructose*	*Maltose Anhydrous*	*Maltose Monohydrate*	*Lactose Anhydrous*	*Lactose Monohydrate*
	0	1	5	10	25						
15	50·5	49·9	47·6	46·1	43·4	49·1	52·2	77·2	81·3	64·9	68·3
16	50·6	50·0	47·6	46·1	43·4	49·2	52·3	77·1	81·2	64·8	68·2
17	50·7	50·1	47·6	46·1	43·4	49·3	52·3	77·0	81·1	64·8	68·2
18	50·8	50·1	47·6	46·1	43·3	49·3	52·4	77·0	81·0	64·7	68·1
19	50·8	50·2	47·6	46·1	43·3	49·4	52·5	76·9	80·9	64·7	68·1
20	50·9	50·2	47·6	46·1	43·2	49·5	52·5	76·8	80·8	64·6	68·0
21	51·0	50·2	47·6	46·1	43·2	49·5	52·6	76·7	80·7	64·6	68·0
22	51·0	50·3	47·6	46·1	43·1	49·6	52·7	76·6	80·6	64·6	68·0
23	51·1	50·3	47·6	46·1	43·0	49·7	52·7	76·5	80·5	64·5	67·9
24	51·2	50·3	47·6	46·1	42·9	49·8	52·8	76·4	80·4	64·5	67·9
25	51·2	50·4	47·6	46·0	42·8	49·8	52·8	76·4	80·4	64·5	67·9
26	51·3	50·4	47·6	46·0	42·8	49·9	52·9	76·3	80·3	64·5	67·9
27	51·4	50·4	47·6	46·0	42·7	49·9	52·9	76·2	80·2	64·4	67·8
28	51·4	50·5	47·7	46·0	42·7	50·0	53·0	76·1	80·1	64·4	67·8
29	51·5	50·5	47·7	46·0	42·6	50·0	53·1	76·0	80·0	64·4	67·8
30	51·5	50·5	47·7	46·0	42·5	50·1	53·2	76·0	80·0	64·4	67·8
31	51·6	50·6	47·7	45·9	42·5	50·2	53·2	75·9	79·9	64·4	67·8
32	51·6	50·6	47·7	45·9	42·4	50·2	53·3	75·9	79·9	64·4	67·8
33	51·7	50·6	47·7	45·9	42·3	50·3	53·3	75·8	79·8	64·4	67·8
34	51·7	50·6	47·7	45·8	42·2	50·3	53·4	75·8	79·8	64·4	67·9
35	51·8	50·7	47·7	45·8	42·2	50·4	53·4	75·7	79·7	64·5	67·9
36	51·8	50·7	47·7	45·8	42·1	50·4	53·5	75·6	79·6	64·5	67·9
37	51·9	50·7	47·7	45·7	42·0	50·5	53·5	75·6	79·6	64·5	67·9
38	51·9	50·7	47·7	45·7	42·0	50·5	53·6	75·5	79·5	64·5	67·9
39	52·0	50·8	47·7	45·7	41·9	50·6	53·6	75·5	79·5	64·5	67·9
40	52·0	50·8	47·7	45·6	41·8	50·6	53·6	75·4	79·4	64·5	67·9
41	52·1	50·8	47·7	45·6	41·8	50·7	53·7	75·4	79·4	64·6	68·0
42	52·1	50·8	47·7	45·6	41·7	50·7	53·7	75·3	79·3	64·6	68·0
43	52·2	50·8	47·7	45·5	41·6	50·8	53·8	75·3	79·3	64·6	68·0
44	52·2	50·9	47·7	45·5	41·5	50·8	53·8	75·2	79·2	64·6	68·0
45	52·3	50·9	47·7	45·4	41·4	50·9	53·9	75·2	79·2	64·7	68·1
46	52·3	50·9	47·7	45·4	41·4	50·9	53·9	75·1	79·1	64·7	68·1
47	52·4	50·9	47·7	45·3	41·3	51·0	53·9	75·1	79·1	64·8	68·2
48	52·4	50·9	47·7	45·3	41·2	51·0	54·0	75·1	79·1	64·8	68·2
49	52·5	51·0	47·7	45·2	41·1	51·0	54·0	75·0	79·0	64·8	68·2
50	52·5	51·0	47·7	45·2	41·0	51·1	54·0	75·0	79·0	64·9	68·3

TABLE A.2—*continued*
FACTORS FOR USE WITH THE LANE–EYNON METHOD
(b) 25 ml Soxhlet reagent

Titre	*Sucrose/100 ml invert sugar* 0	*Sucrose/100 ml invert sugar* 1 g	*Glucose*	*Fructose*	*Maltose Anhydrous*	*Maltose Monohydrate*	*Lactose Anhydrous*	*Lactose Monohydrate*
15	123·6	122·6	120·2	127·4	197·8	208·2	163·9	172·5
16	123·6	122·7	120·2	127·4	197·4	207·8	163·5	172·1
17	123·6	122·7	120·2	127·5	197·0	207·4	163·1	171·7
18	123·7	122·7	120·2	127·5	196·7	207·1	162·8	171·4
19	123·7	122·8	120·3	127·6	196·5	206·8	162·5	171·1
20	123·8	122·8	120·3	127·6	196·2	206·5	162·3	170·9
21	123·8	122·8	120·3	127·7	195·8	206·1	162·0	170·6
22	123·9	122·9	120·4	127·7	195·5	205·8	161·8	170·4
23	123·9	122·9	120·4	127·8	195·1	205·4	161·6	170·2
24	124·0	122·9	120·5	127·8	194·8	205·1	161·5	170·0
25	124·0	123·0	120·5	127·9	194·5	204·8	161·4	169·9
26	124·1	123·0	120·6	127·9	194·2	204·4	161·2	169·7
27	124·1	123·0	120·6	128·0	193·9	204·1	161·0	169·5
28	124·2	123·1	120·7	128·0	193·6	203·8	160·8	169·3
29	124·2	123·1	120·7	128·1	193·3	203·5	160·7	169·2
30	124·3	123·1	120·8	128·1	193·0	203·2	160·6	169·0
31	124·3	123·2	120·8	128·1	192·8	202·9	160·5	168·9
32	124·4	123·2	120·8	128·2	192·5	202·6	160·4	168·8
33	124·4	123·2	120·9	128·2	192·2	202·3	160·2	168·6
34	124·5	123·3	120·9	128·3	191·9	202·0	160·1	168·5
35	124·5	123·3	121·0	128·3	191·7	201·8	160·0	168·4
36	124·6	123·3	121·0	128·4	191·4	201·5	159·8	168·2
37	124·6	123·4	121·1	128·4	191·2	201·2	159·7	168·1
38	124·7	123·4	121·2	128·5	191·0	201·0	159·6	168·0
39	124·7	123·4	121·2	128·5	190·8	200·8	159·5	167·9
40	124·8	123·4	121·2	128·6	190·5	200·5	159·4	167·8
41	124·8	123·5	121·3	128·6	190·3	200·3	159·3	167·7
42	124·9	123·5	121·4	128·6	190·1	200·1	159·2	167·6
43	124·9	123·5	121·4	128·7	189·8	199·8	159·2	167·6
44	125·0	123·6	121·5	128·7	189·6	199·6	159·1	167·5
45	125·0	123·6	121·5	128·8	189·4	199·4	159·0	167·4
46	125·1	123·6	121·6	128·8	189·2	199·2	159·0	167·4
47	125·1	123·7	121·6	128·9	189·0	199·0	158·9	167·3
48	125·2	123·7	121·7	128·9	188·9	198·9	158·8	167·2
49	125·2	123·7	121·7	129·0	188·8	198·7	158·8	167·2
50	125·3	123·8	121·8	129·0	188·7	198·6	158·7	167·1

TABLE A.3

FACTORS[a] FOR USE WITH THE MUNSON AND WALKER GRAVIMETRIC METHOD

Cu_2O	Glucose	Invert sugar	Invert sugar + sucrose 0·4 g total	2 g total	Lactose monohydrate	Lactose:sucrose 1:4	1:12	Maltose monohydrate
10	4·0	4·5	1·6	......	6·3	6·1	......	6·2
12	4·9	5·4	2·5	......	7·5	7·3	......	7·9
14	5·7	6·3	3·4	......	8·8	8·5	......	9·5
16	6·6	7·2	4·3	......	10·0	9·7	......	11·2
18	7·5	8·1	5·2	......	11·3	10·9	......	12·9
20	8·3	8·9	6·1	......	12·5	12·1	......	14·6
22	9·2	9·8	7·0	......	13·8	13·3	......	16·2
24	10·0	10·7	7·9	......	15·0	14·5	......	17·9
26	10·9	11·6	8·8	......	16·3	15·8	......	19·6
28	11·8	12·5	9·7	......	17·6	17·0	......	21·2
30	12·6	13·4	10·7	4·3	18·8	18·2	......	22·9
32	13·5	14·3	11·6	5·2	20·1	19·4	......	24·6
34	14·3	15·2	12·5	6·1	21·4	20·7	......	26·2
36	15·2	16·1	13·4	7·0	22·8	22·0	......	27·9
38	16·1	16·9	14·3	7·9	24·2	23·3	......	29·6
40	16·9	17·8	15·2	8·8	25·5	24·7	......	31·3
42	17·8	18·7	16·1	9·7	26·9	26·0	......	32·9
44	18·7	19·6	17·0	10·7	28·3	27·3	......	34·6
46	19·6	20·5	17·9	11·6	29·6	28·6	......	36·3
48	20·4	21·4	18·8	12·5	31·0	30·0	......	37·9
50	21·3	22·3	19·7	13·4	32·3	31·3	......	39·6
52	22·2	23·2	20·7	14·3	33·7	32·6	......	41·3
54	23·0	24·1	21·6	15·2	35·1	34·0	......	42·9
56	23·9	25·0	22·5	16·2	36·4	35·3	......	44·6
58	24·8	25·9	23·4	17·1	37·8	36·6	......	46·3
60	25·6	26·8	24·3	18·0	39·2	37·9	......	48·0
62	26·5	27·7	25·2	18·9	40·5	39·3	......	49·6
64	27·4	28·6	26·2	19·8	41·9	40·6	......	51·3
66	28·3	29·5	27·1	20·8	43·3	41·9	......	53·0
68	29·2	30·4	28·0	21·7	44·7	43·3	40·7	54·6

[a] Expressed in mg.

TABLE A.3—*continued*

Cu_2O	Glucose	Invert sugar	Invert sugar + sucrose 0·4 g total	2 g total	Lactose monohydrate	Lactose:sucrose 1:4	1:12	Maltose monohydrate
70	30·0	31·3	28·9	22·6	46·0	44·6	41·9	56·3
72	30·9	32·3	29·8	23·5	47·4	45·9	43·1	58·0
74	31·8	33·2	30·8	24·5	48·8	47·3	44·2	59·6
76	32·7	34·1	31·7	25·4	50·1	48·6	45·4	61·3
78	33·6	35·0	32·6	26·3	51·5	49·9	46·6	63·0
80	34·4	35·9	33·5	27·3	52·9	51·3	47·8	64·6
82	35·3	36·8	34·5	28·2	54·2	52·6	49·0	66·3
84	36·2	37·7	35·4	29·1	55·6	53·9	50·1	68·0
86	37·1	38·6	36·3	30·0	57·0	55·3	51·3	69·7
88	38·0	39·5	37·2	31·0	58·4	56·6	52·5	71·3
90	38·9	40·4	38·2	31·9	59·7	57·9	53·7	73·0
92	39·8	41·4	39·1	32·8	61·1	59·3	54·9	74·7
94	40·6	42·3	40·0	33·8	62·5	60·6	56·0	76·3
96	41·5	43·2	41·0	34·7	63·8	61·9	57·2	78·0
98	42·4	44·1	41·9	35·6	65·2	63·3	58·4	79·7
100	43·3	45·0	42·8	36·6	66·6	64·6	59·6	81·3
102	44·2	46·0	43·8	37·5	68·0	66·0	60·8	83·0
104	45·1	46·9	44·7	38·5	69·3	67·3	62·0	84·7
106	46·0	47·8	45·6	39·4	70·7	68·6	63·2	86·3
108	46·9	48·7	46·6	40·3	72·1	70·0	64·4	88·0
110	47·8	49·6	47·5	41·3	73·5	71·3	65·6	89·7
112	48·7	50·6	48·4	42·2	74·8	72·6	66·7	91·3
114	49·6	51·5	49·4	43·2	76·2	74·0	67·9	93·0
116	50·5	52·4	50·3	44·1	77·6	75·3	69·1	94·7
118	51·4	53·3	51·2	45·0	79·0	76·7	70·3	96·4
120	52·3	54·3	52·2	46·0	80·3	78·0	71·5	98·0
122	53·2	55·2	53·1	46·9	81·7	79·3	72·7	99·7
124	54·1	56·1	54·1	47·9	83·1	80·7	73·9	101·4
126	55·0	57·0	55·0	48·8	84·5	82·0	75·1	103·0
128	55·9	58·0	55·9	49·8	85·8	83·4	76·3	104·7
130	56·8	58·9	56·9	50·7	87·2	84·7	77·5	106·4
132	57·7	59·8	57·8	51·7	88·6	86·0	78·7	108·0
134	58·6	60·8	58·8	52·6	90·0	87·4	79·7	109·7
136	59·5	61·7	59·7	53·6	91·3	88·7	81·1	111·4
138	60·4	62·6	60·7	54·5	92·7	90·1	82·3	113·0

TABLE A.3—*continued*

Cu_2O	Glucose	Invert sugar	Invert sugar + sucrose 0·4 g total	2 g total	Lactose monohydrate	Lactose:sucrose 1:4	1:12	Maltose monohydrate
140	61·3	63·6	61·6	55·5	94·1	91·4	83·5	114·7
142	62·2	64·5	62·6	56·4	95·5	92·8	84·7	116·4
144	63·1	65·4	63·5	57·4	96·8	94·1	85·9	118·0
146	64·0	66·4	64·5	58·3	98·2	95·4	87·1	119·7
148	65·0	67·3	65·4	59·3	99·6	96·8	88·3	121·4
150	65·9	68·3	66·4	60·2	101·0	98·1	89·5	123·0
152	66·8	69·2	67·3	61·2	102·3	99·5	90·8	124·7
154	67·7	70·1	68·3	62·1	103·7	100·8	92·0	126·4
156	68·6	71·1	69·2	63·1	105·1	102·2	93·2	128·0
158	69·5	72·0	70·2	64·1	106·5	103·5	94·4	129·7
160	70·4	73·0	71·2	65·0	107·9	104·8	95·6	131·4
162	71·4	73·9	72·1	66·0	109·2	106·2	96·8	133·0
164	72·3	74·9	73·1	66·9	110·6	107·5	98·0	134·7
166	73·2	75·8	74·0	67·9	112·0	108·9	99·2	136·4
168	74·1	76·8	75·0	68·8	113·4	110·2	100·4	138·0
170	75·1	77·7	76·0	69·8	114·8	111·6	101·6	139·7
172	76·0	78·7	76·9	70·8	116·1	112·9	102·8	141·4
174	76·9	79·6	77·9	71·7	117·5	114·3	104·1	143·0
176	77·8	80·6	78·8	72·7	118·9	115·6	105·3	144·7
178	78·8	81·5	79·8	73·7	120·3	117·0	106·5	146·4
180	79·7	82·5	80·8	74·6	121·6	118·3	107·7	148·0
182	80·6	83·4	81·7	75·6	123·1	119·7	108·9	149·7
184	81·5	84·4	82·7	76·6	124·3	121·0	110·1	151·4
186	82·5	85·3	83·7	77·6	125·8	122·4	111·3	153·0
188	83·4	86·3	84·6	78·5	127·2	123·7	112·5	154·7
190	84·3	87·2	85·6	79·5	128·5	125·1	113·8	156·4
192	85·3	88·2	86·6	80·5	129·9	126·4	115·0	158·0
194	86·2	89·2	87·6	81·4	131·3	127·8	116·2	159·7
196	87·1	90·1	88·5	82·4	132·7	129·2	117·4	161·4
198	88·1	91·1	89·5	83·4	134·1	130·5	118·6	163·0
200	89·0	92·0	90·5	84·4	135·4	131·9	119·8	164·7
202	89·9	93·0	91·4	85·3	136·8	133·2	121·0	166·4
204	90·9	94·0	92·4	86·3	138·2	134·6	122·3	168·0
206	91·8	94·9	93·4	87·3	139·6	135·9	123·5	169·7
208	92·8	95·9	94·4	88·3	141·0	137·3	124·7	171·4

Appendix

TABLE A.3—*continued*

Cu_2O	Glucose	Invert sugar	Invert sugar + sucrose 0·4 g total	2 g total	Lactose monohydrate	Lactose:sucrose 1:4	1:12	Maltose monohydrate
210	93·7	96·9	95·4	89·2	142·3	138·6	126·0	173·0
212	94·6	97·8	96·3	90·2	143·7	140·0	127·2	174·7
214	95·6	98·8	97·3	91·2	145·1	141·4	128·4	176·4
216	96·5	99·8	98·3	92·2	146·5	142·7	129·6	178·0
218	97·5	100·8	99·3	93·2	147·9	144·1	130·9	179·7
220	98·4	101·7	100·3	94·2	149·3	145·4	132·1	181·4
222	99·4	102·7	101·2	95·1	150·7	146·8	133·3	183·0
224	100·3	103·7	102·2	96·1	152·0	148·1	134·5	184·7
226	101·3	104·6	103·2	97·1	153·4	149·5	135·8	186·4
228	102·2	105·6	104·2	98·1	154·8	150·8	137·0	188·0
230	103·2	106·6	105·2	99·1	156·2	152·2	138·2	189·7
232	104·1	107·6	106·2	100·1	157·6	153·6	139·4	191·3
234	105·1	108·6	107·2	101·1	159·0	154·9	140·7	193·0
236	106·0	109·5	108·2	102·1	160·3	156·3	141·9	194·7
238	107·0	110·5	109·2	103·1	161·7	157·6	143·2	196·3
240	108·0	111·5	110·1	104·0	163·1	159·0	144·4	198·0
242	108·9	112·5	111·1	105·0	164·5	160·3	145·6	199·7
244	109·9	113·5	112·1	106·0	165·9	161·7	146·9	201·3
246	110·8	114·5	113·1	107·0	167·3	163·1	148·1	203·0
248	111·8	115·4	114·1	108·0	168·7	164·4	149·3	204·7
250	112·8	116·4	115·1	109·0	170·1	165·8	150·6	206·3
252	113·7	117·4	116·1	110·0	171·5	167·2	151·8	208·0
254	114·7	118·4	117·1	111·0	172·8	168·5	153·1	209·7
256	115·7	119·4	118·1	112·0	174·2	169·9	154·3	211·3
258	116·6	120·4	119·1	113·0	175·6	171·3	155·5	213·0
260	117·6	121·4	120·1	114·0	177·0	172·6	156·8	214·7
262	118·6	122·4	121·1	115·0	178·4	174·0	158·0	216·3
264	119·5	123·4	122·1	116·0	179·8	175·3	159·3	218·0
266	120·5	124·4	123·1	117·0	181·2	176·7	160·5	219·7
268	121·5	125·4	124·1	118·0	182·6	178·1	161·8	221·3
270	122·5	126·4	125·1	119·0	184·0	179·4	163·0	223·0
272	123·4	127·4	126·2	120·0	185·3	180·8	164·3	224·6
274	124·4	128·4	127·2	121·1	186·7	182·2	165·5	226·3
276	125·4	129·4	128·2	122·1	188·1	183·5	166·8	228·0
278	126·4	130·4	129·2	123·1	189·5	184·9	168·0	229·6

TABLE A.3—*continued*

Cu_2O	Glucose	Invert sugar	Invert sugar + sucrose 0·4 g total	2 g total	Lactose monohydrate	Lactose:sucrose 1:4	1:12	Maltose monohydrate
280	127·3	131·4	130·2	124·1	190·9	186·3	169·3	231·3
282	128·3	132·4	131·2	125·1	192·3	187·6	170·5	233·0
284	129·3	133·4	132·2	126·1	193·7	189·0	171·8	234·6
286	130·3	134·4	133·2	127·1	195·1	190·4	173·0	236·3
288	131·3	135·4	134·3	128·1	196·5	191·7	174·3	238·0
290	132·3	136·4	135·3	129·2	197·8	193·1	175·5	239·6
292	133·2	137·4	136·3	130·2	199·2	194·4	176·8	241·3
294	134·2	138·4	137·3	131·2	200·6	195·8	178·1	242·9
296	135·2	139·4	138·3	132·2	202·0	197·2	179·3	244·6
298	136·2	140·5	139·4	133·2	203·4	198·6	180·6	246·3
300	137·2	141·5	140·4	134·2	204·8	199·9	181·8	247·9
302	138·2	142·5	141·4	135·3	206·2	201·3	183·1	249·6
304	139·2	143·5	142·4	136·3	207·6	202·7	184·4	251·3
306	140·2	144·5	143·4	137·3	209·0	204·0	185·6	252·9
308	141·2	145·5	144·5	138·3	210·4	205·4	186·9	254·6
310	142·2	146·6	145·5	139·4	211·8	206·8	188·1	256·3
312	143·2	147·6	146·5	140·4	213·2	208·1	189·4	257·9
314	144·2	148·6	147·6	141·4	214·6	209·5	190·7	259·6
316	145·2	149·6	148·6	142·4	216·0	210·9	191·9	261·2
318	146·2	150·7	149·6	143·5	217·3	212·2	193·2	262·9
320	147·2	151·7	150·7	144·5	218·7	213·6	194·4	264·6
322	148·2	152·7	151·7	145·5	220·1	215·0	195·7	266·2
324	149·2	153·7	152·7	146·6	221·5	216·4	197·0	267·9
326	150·2	154·8	153·8	147·6	222·9	217·7	198·2	269·6
328	151·2	155·8	154·8	148·6	224·3	219·1	199·5	271·2
330	152·2	156·8	155·8	149·7	225·7	220·5	200·8	272·9
332	153·2	157·9	156·9	150·7	227·1	221·8	202·0	274·6
334	154·2	158·9	157·9	151·7	228·5	223·2	203·3	276·2
336	155·2	159·9	159·0	152·8	229·9	224·6	204·6	277·9
338	156·3	161·0	160·0	153·8	231·3	226·0	205·9	279·5
340	157·3	162·0	161·0	154·8	232·7	227·4	207·1	281·2
342	158·3	163·1	162·1	155·9	234·1	228·7	208·4	282·9
344	159·3	164·1	163·1	156·9	235·5	230·1	209·7	284·5
346	160·3	165·1	164·2	158·0	236·9	231·5	211·0	286·2
348	161·4	166·2	165·2	159·0	238·3	232·9	212·2	287·9

TABLE A.3—*continued*

Cu_2O	Glucose	Invert sugar	Invert sugar + sucrose 0·4 g total	2 g total	Lactose monohydrate	Lactose:sucrose 1:4	1:12	Maltose monohydrate
350	162·4	167·2	166·3	160·1	239·7	234·3	213·5	289·5
352	163·4	168·3	167·3	161·1	241·1	235·6	214·8	291·2
354	164·4	169·3	168·4	162·2	242·5	237·0	216·1	292·8
356	165·4	170·4	169·4	163·2	243·9	238·4	217·3	294·5
358	166·5	171·4	170·5	164·3	245·3	239·8	218·6	296·2
360	167·5	172·5	171·5	165·3	246·7	241·2	219·9	297·8
362	168·5	173·5	172·6	166·4	248·1	242·5	221·2	299·5
364	169·6	174·6	173·7	167·4	249·5	243·9	222·5	301·2
366	170·6	175·6	174·7	168·5	250·9	245·3	223·7	302·8
368	171·6	176·7	175·8	169·5	252·3	246·7	225·0	304·5
370	172·7	177·7	176·8	170·6	253·7	248·1	226·3	306·1
372	173·7	178·8	177·9	171·6	255·1	249·5	227·6	307·8
374	174·7	179·8	179·0	172·7	256·5	250·9	228·9	309·5
376	175·8	180·9	180·0	173·7	257·9	252·2	230·2	311·1
378	176·8	182·0	181·1	174·8	259·3	253·6	231·5	312·8
380	177·9	183·0	182·1	175·9	260·7	255·0	232·8	314·5
382	178·9	184·1	183·2	176·9	262·1	256·4	234·1	316·1
384	180·0	185·2	184·3	178·0	263·5	257·8	235·4	317·8
386	181·0	186·2	185·4	179·1	264·9	259·2	236·6	319·4
388	182·0	187·3	186·4	180·1	266·5	260·5	237·9	321·1
390	183·1	188·4	187·5	181·2	267·7	261·9	239·2	322·8
392	184·1	189·4	188·6	182·3	269·1	263·3	240·5	324·4
394	185·2	190·5	189·7	183·3	270·5	264·7	241·8	326·1
396	186·2	191·6	190·7	184·4	271·9	266·1	243·1	327·7
398	187·3	192·7	191·8	185·5	273·3	267·5	244·4	329·4
400	188·4	193·7	192·9	186·5	274·7	268·9	245·7	331·1
402	189·4	194·8	194·0	187·6	276·1	270·3	247·0	332·7
404	190·5	195·9	195·0	188·7	277·5	271·7	248·3	334·4
406	191·5	197·0	196·1	189·8	278·9	273·0	249·6	336·0
408	192·6	198·1	197·2	190·8	280·3	274·4	251·0	337·7
410	193·7	199·1	198·3	191·9	281·7	275·8	252·3	339·4
412	194·7	200·2	199·4	193·0	283·2	277·2	253·6	341·0
414	195·8	201·3	200·5	194·1	284·6	278·6	254·9	342·7
416	196·8	202·4	201·6	195·2	286·0	280·0	256·2	344·4
418	197·9	203·5	202·6	196·2	287·4	281·4	257·5	346·0

TABLE A.3—*continued*

Cu_2O	Glucose	Invert sugar	Invert sugar + sucrose 0·4 g total	2 g total	Lactose monohydrate	Lactose:sucrose 1:4	1:12	Maltose monohydrate
420	199·0	204·6	203·7	197·3	288·8	282·8	258·8	347·7
422	200·1	205·7	204·8	198·4	290·2	284·2	260·1	349·3
424	201·1	206·7	205·9	199·5	291·6	285·6	261·4	351·0
426	202·2	207·8	207·0	200·6	293·0	287·0	262·7	352·7
428	203·3	208·9	208·1	201·7	294·4	288·4	264·0	354·3
430	204·4	210·0	209·2	202·7	295·8	289·8	265·4	356·0
432	205·5	211·1	210·3	203·8	297·2	291·2	266·6	357·6
434	206·5	212·2	211·4	204·9	298·6	292·6	268·0	359·3
436	207·6	213·3	212·5	206·0	300·0	294·0	269·3	361·0
438	208·7	214·4	213·6	207·1	301·4	295·4	270·6	362·6
440	209·8	215·5	214·7	208·2	302·8	296·8	272·0	364·3
442	210·9	216·6	215·8	209·3	304·2	298·2	273·3	365·9
444	212·0	217·8	216·9	210·4	305·6	299·6	274·6	367·6
446	213·1	218·9	218·0	211·5	307·0	301·0	275·9	369·3
448	214·1	220·0	219·1	212·6	308·4	302·4	277·2	370·9
450	215·2	221·1	220·2	213·7	309·9	303·8	278·6	372·6
452	216·3	222·2	221·4	214·8	311·3	305·2	279·9	374·2
454	217·4	223·3	222·5	215·9	312·7	306·6	281·2	375·9
456	218·5	224·4	223·6	217·0	314·1	308·0	282·5	377·6
458	219·6	225·5	224·7	218·1	315·5	309·4	283·9	379·2
460	220·7	226·7	225·8	219·2	316·9	310·8	285·2	380·9
462	221·8	227·8	226·9	220·3	318·3	312·2	286·5	382·5
464	222·9	228·9	228·1	221·4	319·7	313·6	287·8	384·2
466	224·0	230·0	229·2	222·5	321·1	315·0	289·2	385·9
468	225·1	231·2	230·3	223·7	322·5	316·4	290·5	387·5
470	226·2	232·3	231·4	224·8	323·9	317·7	291·8	389·2
472	227·4	233·4	232·5	225·9	325·3	319·1	293·2	390·8
474	228·3	234·5	233·7	227·0	326·8	320·5	294·5	392·5
476	229·6	235·7	234·8	228·1	328·2	321·9	295·8	394·2
478	230·7	236·8	235·9	229·2	329·6	323·3	297·1	395·8
480	231·8	237·9	237·1	230·3	331·0	324·7	298·5	397·5
482	232·9	239·1	238·2	231·5	332·4	326·1	299·8	399·1
484	234·1	240·2	239·3	232·6	333·8	327·5	301·1	400·8
486	235·2	241·4	240·5	233·7	335·2	328·9	302·5	402·4
488	236·3	242·5	241·6	234·8	336·6	330·3	303·8	404·1
490	237·4	243·6	242·7	236·0	338·0	331·7	305·1	405·8

TABLE A.4
HAMMOND FACTORS[a] FOR USE WITH THE MUNSON AND WALKER VOLUMETRIC METHOD

Cu	*Glucose*	*Fructose*	*Lactose monohydrate*	*Invert sugar*	*Invert sugar + sucrose*		
					0·3 g total	*0·4 g total*	*2 g total*
10	4·6	5·1	7·7	5·2	3·2	2·9	......
12	5·6	6·1	9·3	6·2	4·2	3·9	......
14	6·5	7·2	10·8	7·2	5·3	4·9	......
16	7·5	8·3	12·3	8·2	6·3	5·9	......
18	8·5	9·3	13·8	9·2	7·3	6·9	......
20	9·4	10·4	15·4	10·2	8·3	7·9	1·9
22	10·4	11·5	16·9	11·2	9·3	8·9	2·9
24	11·4	12·5	18·4	12·3	10·4	10·0	3·9
26	12·3	13·6	19·9	13·3	11·4	11·0	4·9
28	13·3	14·7	21·5	14·3	12·4	12·0	6·0
30	14·3	15·8	23·0	15·3	13·4	13·0	7·0
32	15·3	16·8	24·5	16·3	14·5	14·1	8·0
34	16·2	17·9	26·1	17·3	15·5	15·1	9·0
36	17·2	19·0	27·6	18·3	16·5	16·1	10·1
38	18·2	20·1	29·1	19·4	17·6	17·1	11·1
40	19·2	21·1	30·6	20·4	18·6	18·2	12·1
42	20·1	22·2	32·2	21·4	19·6	19·2	13·1
44	21·1	23·3	33·7	22·4	20·7	20·2	14·2
46	22·1	24·4	35·2	23·5	21·7	21·3	15·2
48	23·1	25·4	36·8	24·5	22·7	22·3	16·2
50	24·1	26·5	38·3	25·5	23·8	23·3	17·3
52	25·1	27·6	39·8	26·5	24·8	24·3	18·3
54	26·1	28·7	41·4	27·6	25·8	25·4	19·3
56	27·0	29·8	42·9	28·6	26·9	26·4	20·4
58	28·0	30·9	44·4	29·6	27·9	27·5	21·4
60	29·0	31·9	46·0	30·6	28·9	28·5	22·5
62	30·0	33·0	47·5	31·7	30·0	29·5	23·5
64	31·0	34·1	49·0	32·7	31·0	30·6	24·5
66	32·0	35·2	50·6	33·7	32·1	31·6	25·6
68	33·0	36·3	52·1	34·8	33·1	32·7	26·6

[a] Expressed in mg.

TABLE A.4—*continued*

Cu	*Glucose*	*Fructose*	*Lactose monohydrate*	*Invert sugar*	Invert sugar + sucrose		
					0·3 g total	*0·4 g total*	*2 g total*
70	34·0	37·4	53·6	35·8	34·2	33·7	27·7
72	35·0	38·5	55·2	36·8	35·2	34·7	28·7
74	36·0	39·6	56·7	37·9	36·3	35·8	29·8
76	37·0	40·7	58·2	38·9	37·3	36·8	30·8
78	38·0	41·7	59·8	40·0	38·4	37·9	31·9
80	39·0	42·8	61·3	41·0	39·4	38·9	32·9
82	40·0	43·9	62·8	42·0	40·5	40·0	34·0
84	41·0	45·0	64·4	43·1	41·5	41·0	35·0
86	42·0	46·1	65·9	44·1	42·6	42·1	36·1
88	43·0	47·2	67·4	45·2	43·6	43·1	37·1
90	44·0	48·3	69·0	46·2	44·7	44·2	38·2
92	45·0	49·4	70·5	47·3	45·7	45·2	39·2
94	46·0	50·5	72·1	48·3	46·8	46·3	40·3
96	47·0	51·6	73·6	49·4	47·8	47·4	41·3
98	48·0	52·7	75·1	50·4	48·9	48·4	42·4
100	49·0	53·8	76·7	51·5	50·0	49·5	43·5
102	50·0	54·9	78·2	52·5	51·0	50·5	44·5
104	51·1	56·0	79·7	53·6	52·1	51·6	45·6
106	52·1	57·1	81·3	54·6	53·1	52·7	46·7
108	53·1	58·2	82·8	55·7	54·2	53·7	47·7
110	54·1	59·3	84·4	56·7	55·3	54·8	48·8
112	55·1	60·4	85·9	57·8	56·3	55·8	49·9
114	56·1	61·6	87·4	58·9	57·4	56·9	50·9
116	57·2	62·7	89·0	59·9	58·5	58·0	52·0
118	58·2	63·8	90·5	61·0	59·5	59·0	53·1
120	59·2	64·9	92·1	62·0	60·6	60·1	54·1
122	60·2	66·0	93·6	63·1	61·7	61·2	55·2
124	61·3	67·1	95·2	64·2	62·8	62·3	56·3
126	62·3	68·2	96·7	65·2	63·8	63·3	57·4
128	63·3	69·3	98·2	66·3	64·9	64·4	58·4

Appendix

TABLE A.4—*continued*

Cu	Glucose	Fructose	Lactose monohydrate	Invert sugar	Invert sugar + sucrose		
					0·3 g total	0·4 g total	2 g total
130	64·3	70·4	99·8	67·4	66·0	65·5	59·5
132	65·4	71·6	101·3	68·4	67·1	66·6	60·6
134	66·4	72·7	102·9	69·5	68·1	67·6	61·7
136	67·4	73·8	104·4	70·6	69·2	68·7	62·8
138	68·5	74·9	106·0	71·6	70·3	69·8	63·9
140	69·5	76·0	107·5	72·7	71·4	70·9	64·9
142	70·5	77·1	109·0	73·8	72·5	72·0	66·0
144	71·6	78·3	110·6	74·9	73·5	73·0	67·1
146	72·6	79·4	112·1	75·9	74·6	74·1	68·2
148	73·7	80·5	113·7	77·0	75·7	75·2	69·3
150	74·7	81·6	115·2	78·1	76·8	76·3	70·4
152	75·7	82·8	116·8	79·2	77·9	77·4	71·5
154	76·8	83·9	118·3	80·3	79·0	78·5	72·6
156	77·8	85·0	119·9	81·3	80·1	79·6	73·7
158	78·9	86·1	121·4	82·4	81·2	80·6	74·8
160	79·9	87·3	122·9	83·5	82·2	81·7	75·9
162	81·0	88·4	124·5	84·6	83·3	82·8	77·0
164	82·0	89·5	126·0	85·7	84·4	83·9	78·1
166	83·1	90·6	127·6	86·8	85·5	85·0	79·2
168	84·1	91·8	129·1	87·8	86·6	86·1	80·3
170	85·2	92·9	130·7	88·9	87·7	87·2	81·4
172	86·2	94·0	132·2	90·0	88·8	88·3	82·5
174	87·3	95·2	133·8	91·1	89·9	89·4	83·6
176	88·3	96·3	135·3	92·2	91·0	90·5	84·7
178	89·4	97·4	136·9	93·3	92·1	91·6	85·8
180	90·4	98·6	138·4	94·4	93·2	92·7	86·9
182	91·5	99·7	140·0	95·5	94·3	93·8	88·0
184	92·6	100·9	141·5	96·6	95·4	94·9	89·1
186	93·6	102·0	143·1	97·7	96·5	96·0	90·2
188	94·7	103·1	144·6	98·8	97·6	97·1	91·3

TABLE A.4—*continued*

Cu	*Glucose*	*Fructose*	*Lactose monohydrate*	*Invert sugar*	*Invert sugar + sucrose*		
					0·3 g total	*0·4 g total*	*2 g total*
190	95·7	104·3	146·2	99·9	98·7	98·2	92·4
192	96·8	105·4	147·7	101·0	99·9	99·4	93·6
194	97·9	106·6	149·3	102·1	101·0	100·5	94·7
196	98·9	107·7	150·8	103·2	102·1	101·6	95·8
198	100·0	108·8	152·4	104·3	103·2	102·7	96·9
200	101·1	110·0	153·9	105·4	104·3	103·8	98·0
202	102·2	111·1	155·5	106·5	105·4	104·9	99·2
204	103·2	112·3	157·0	107·6	106·5	106·0	100·3
206	104·3	113·4	158·6	108·7	107·6	107·2	101·4
208	105·4	114·6	160·2	109·8	108·8	108·3	102·5
210	106·5	115·7	161·7	110·9	109·9	109·4	103·7
212	107·5	116·9	163·3	112·1	111·0	110·5	104·8
214	108·6	118·0	164·8	113·2	112·1	111·6	105·9
216	109·7	119·2	166·4	114·3	113·2	112·8	107·1
218	110·8	120·3	167·9	115·4	114·4	113·9	108·2
220	111·9	121·5	169·5	116·5	115·5	115·0	109·3
222	112·9	122·6	171·0	117·6	116·6	116·1	110·5
224	114·0	123·8	172·6	118·8	117·7	117·3	111·6
226	115·1	125·0	174·2	119·9	118·9	118·4	112·7
228	116·2	126·1	175·7	121·0	120·0	119·5	113·9
230	117·3	127·3	177·3	122·1	121·1	120·7	115·0
232	118·4	128·4	178·8	123·3	122·3	121·8	116·2
234	119·5	129·6	180·4	124·4	123·4	122·9	117·3
236	120·6	130·8	181·9	125·5	124·5	124·1	118·4
238	121·7	131·9	183·5	126·6	125·7	125·2	119·6
240	122·7	133·1	185·1	127·8	126·8	126·3	120·7
242	123·8	134·2	186·6	128·9	127·9	127·5	121·9
244	124·9	135·4	188·2	130·0	129·1	128·6	123·0
246	126·0	136·6	189·7	131·2	130·2	129·8	124·2
248	127·1	137·7	191·3	132·3	131·3	130·9	125·3

TABLE A.4—*continued*

Cu	*Glucose*	*Fructose*	*Lactose monohydrate*	*Invert sugar*	*Invert sugar + sucrose*		
					0·3 g total	*0·4 g total*	*2 g total*
250	128·2	138·9	192·9	133·4	132·5	132·0	126·5
252	129·3	140·1	194·4	134·6	133·6	133·2	127·6
254	130·4	141·3	196·0	135·7	134·8	134·3	128·8
256	131·6	142·4	197·5	136·8	135·9	135·5	130·0
258	132·7	143·6	199·1	138·0	137·1	136·6	131·1
260	133·8	144·8	200·7	139·1	138·2	137·8	132·3
262	134·9	145·9	202·2	140·3	139·4	138·9	133·4
264	136·0	147·1	203·8	141·4	140·5	140·1	134·6
266	137·1	148·3	205·3	142·6	141·7	141·2	135·8
268	138·2	149·5	206·9	143·7	142·8	142·4	136·9
270	139·3	150·6	208·5	144·8	144·0	143·5	138·1
272	140·4	151·8	210·0	146·0	145·1	144·7	139·3
274	141·6	153·0	211·6	147·1	146·3	145·9	140·4
276	142·7	154·2	213·2	148·3	147·4	147·0	141·6
278	143·8	155·4	214·7	149·4	148·6	148·2	142·8
280	144·9	156·5	216·3	150·6	149·7	149·3	143·9
282	146·0	157·7	217·9	151·8	150·9	150·5	145·1
284	147·2	158·9	219·4	152·9	152·1	151·7	146·3
286	148·3	160·1	221·0	154·1	153·2	152·8	147·5
288	149·4	161·3	222·6	155·2	154·4	154·0	148·6
290	150·5	162·5	224·1	156·4	155·5	155·2	149·8
292	151·7	163·7	225·7	157·5	156·7	156·3	151·0
294	152·8	164·9	227·3	158·7	157·9	157·5	152·2
296	153·9	166·0	228·8	159·9	159·0	158·7	153·4
298	155·1	167·2	230·4	161·0	160·2	159·9	154·6
300	156·2	168·4	232·0	162·2	161·4	161·0	155·7
302	157·3	169·6	233·5	163·4	162·5	162·2	156·9
304	158·5	170·8	235·1	164·5	163·7	163·4	158·1
306	159·6	172·0	236·7	165·7	164·9	164·6	159·3
308	160·7	173·2	238·2	166·9	166·1	165·7	160·5

TABLE A.4—*continued*

Cu	*Glucose*	*Fructose*	*Lactose monohydrate*	*Invert sugar*	*Invert sugar + sucrose* 0·3 g total	0·4 g total	2 g total
310	161·9	174·4	239·8	168·0	167·2	166·9	161·7
312	163·0	175·6	241·4	169·2	168·4	168·1	162·9
314	164·2	176·8	243·0	170·4	169·6	169·3	164·1
316	165·3	178·0	244·5	171·6	170·8	170·5	165·3
318	166·5	179·2	246·1	172·8	172·0	171·7	166·5
320	167·6	180·4	247·7	173·9	173·1	172·8	167·7
322	168·8	181·6	249·2	175·1	174·3	174·0	168·9
324	169·9	182·8	250·8	176·3	175·5	175·2	170·1
326	171·1	184·0	252·4	177·5	176·7	176·4	171·3
328	172·2	185·2	253·9	178·7	177·9	177·6	172·5
330	173·4	186·4	255·5	179·8	179·1	178·8	173·7
332	174·5	187·6	257·1	181·0	180·3	180·0	174·9
334	175·7	188·8	258·7	182·2	181·5	181·2	176·1
336	176·8	190·1	260·2	183·4	182·6	182·4	177·3
338	178·0	191·3	261·8	184·6	183·8	183·6	178·6
340	179·2	192·5	263·4	185·8	185·0	184·8	179·8
342	180·3	193·7	265·0	187·0	186·2	186·0	181·0
344	181·5	194·9	266·6	188·2	187·4	187·2	182·2
346	182·7	196·1	268·1	189·4	188·6	188·4	183·4
348	183·8	197·3	269·7	190·6	189·8	189·6	184·6
350	185·0	198·5	271·3	191·8	191·0	190·8	185·9
352	186·2	199·8	272·9	193·0	192·2	192·0	187·1
354	187·3	201·0	274·4	194·2	193·4	193·2	188·3
356	188·5	202·2	276·0	195·4	194·6	194·4	189·5
358	189·7	203·4	277·6	196·6	195·8	195·7	190·8
360	190·9	204·7	279·2	197·8	197·1	196·9	192·0
362	192·0	205·9	280·8	199·0	198·3	198·1	193·2
364	193·2	207·1	282·4	200·2	199·5	199·3	194·5
366	194·4	208·3	284·0	201·4	200·7	200·5	195·7
368	195·6	209·6	285·6	202·6	201·9	201·7	196·9

TABLE A.4 —*continued*

Cu	Glucose	Fructose	Lactose monohydrate	Invert sugar	Invert sugar + sucrose 0·3 g total	Invert sugar + sucrose 0·4 g total	Invert sugar + sucrose 2 g total
370	196·8	210·8	287·1	203·8	203·1	203·0	198·2
372	198·0	212·0	288·7	205·0	204·3	204·2	199·4
374	199·1	213·3	290·3	206·3	205·6	205·4	200·7
376	200·3	214·5	291·9	207·5	206·8	206·6	201·9
378	201·5	215·7	293·5	208·7	208·0	207·9	203·1
380	202·7	217·0	295·0	209·9	209·2	209·1	204·4
382	203·9	218·2	296·6	211·1	210·4	210·3	205·6
384	205·1	219·5	298·2	212·4	211·7	211·6	206·9
386	206·3	220·7	299·8	213·6	212·9	212·8	208·1
388	207·5	221·9	301·4	214·8	214·1	214·0	209·4
390	208·7	223·2	303·0	216·0	215·4	215·3	210·6
392	209·9	224·4	304·6	217·3	216·6	216·5	211·9
394	211·1	225·7	306·2	218·5	217·8	217·8	213·2
396	212·3	226·9	307·8	219·8	219·1	219·0	214·4
398	213·5	228·2	309·4	221·0	220·3	220·3	215·7
400	214·7	229·4	311·0	222·2	221·5	221·5	217·0
402	215·9	230·7	312·6	223·5	222·8	222·8	218·2
404	217·1	232·0	314·2	224·7	224·0	224·0	219·5
406	218·4	233·2	315·9	226·0	225·3	225·3	220·8
408	219·6	234·5	317·5	227·2	226·6	226·5	222·0
410	220·8	235·8	319·1	228·5	227·8	227·8	223·3
412	222·0	237·1	320·7	229·7	229·1	229·1	224·6
414	223·3	238·4	322·4	231·0	230·4	230·4	225·9
416	224·5	239·7	324·0	232·3	231·6	231·7	227·2
418	225·7	241·0	325·7	233·6	232·9	232·9	228·5
420	227·0	242·2	327·4	234·8	234·2	234·2	229·8
422	228·2	243·6	329·1	236·1	235·5	235·5	231·1
424	229·5	244·9	330·8	237·5	236·8	236·9	232·4
426	230·7	246·3	332·6	238·8	238·2	238·2	233·8
428	232·0	247·8	334·4	240·2	239·5	239·6	235·1
430	233·3	249·2	336·3	241·5	240·9	241·0	236·5
432	234·7	250·8	338·3	243·0	242·4	242·5	238·0
434	236·1	252·7	340·7	244·7	244·1	244·2	239·6

Bibliography and References

BIBLIOGRAPHY

Association of Official Analytical Chemists, *Official Methods of Analysis*, AOAC, Washington D.C., 1975.

Dawson, R. M. C., Elliott, D. C., Elliott, W. H. and Jones, K. M. *Data for Biochemical Research*, Oxford University Press, London, 2nd edition, 1969.

Peach, K. and Tracey, M. J. (eds). *Modern Methods of Plant Analysis*, Springer-Verlag, Berlin, Gottingen and Heidelberg, 1955.

Pigman, W. and Horton, D. (eds). *The Carbohydrates, Chemistry and Biochemistry*, Academic Press, New York and London, 2nd edition, Vols IIA, IIB, 1970; Vol. IA, 1972.

Whistler, R. L. and Wolfrom, M. L. (eds). *Methods in Carbohydrate Chemistry*, Academic Press, New York, Vol. I, 1962; Vol. V, 1965.

REFERENCES

1. Albaum, H. G. and Umbreit, W. W. (1947). *J. Biol. Chem.*, **167**, 369–376.
2. Albersheim, P. (1965). In *Plant Biochemistry* (Bonner, J. and Vanner, J. E., eds), pp. 298–321, Academic Press, New York and London.
3. Albersheim, P., Nevins, D. J., English, P. D. and Karr, A. (1967). *Carbohyd. Res.*, **5**, 340–345.
4. Association of Official Analytical Chemists (1975). *Official Methods of Analysis*, 12th edition, AOAC, Washington, D.C.
5. Aspinall, G. O. (1969). *Adv. Carbohyd. Chem.*, **24**, 333–379.
6. Aspinall, G. O. (1970). In *The Carbohydrates, Chemistry and Biochemistry* (Pigman, W. and Horton, D., eds), vol. IIB, pp. 515–536, Academic Press, New York & London.
7. Bahl, R. K. (1971). *Analyst*, **96**, 88–92.
8. Bahl, R. K. (1972). *Analyst*, **97**, 559–561.
9. Baker, G. L. (1948). *Adv. Food Res.*, **1**, 395–427.
10. Balazs, E. A., Bernstein, K. O., Karossa, J. and Swann, D. A. (1965). *Anal. Biochem.*, **12**, 547–558.
11. Barton, R. R. (1966). *Anal. Biochem.*, **14**, 258–260.
12. Bath, I. H. (1960). *J. Sci. Food Agric.*, **11**, 560–566.
13. Bergmeyer, H. V. (1970). In *Methoden der Enzymatischen Analyse* (Bergmeyer, H. V. ed.), 2nd edition, vol. II, p. 1163, Verlag Chemie, Weinheim.

14. Birch, G. G. and Green, L. F. (eds) (1973). *Molecular Structure and Function of Food Carbohydrate*, Applied Science Publishers Ltd, London.
15. Bishop, C. T. (1964). *Adv. Carbohyd. Chem.*, **19**, 95–147.
16. Bitter, T. and Muir, H. M. (1962). *Anal. Biochem.*, **4**, 330–334.
17. Blake, J. D. and Richards, G. N. (1971). *Carbohyd. Res.*, **17**, 253–268.
18. Blake, J. D., Murphy, P. T. and Richards, G. N. (1971). *Carbohy. Res.*, **16**, 49–57.
19. Bociek, S. M. and Welti, D. (1975). *Carbohyd. Res.*, **42**, 217–226.
20. Boehringer (1971). *Food Analysis Methods*, Boehringer Mannheim GmbH.
21. Bolton, W. (1960). *Analyst*, **85**, 189–192.
22. Brown, E. G. and Hayes, T. J. (1952). *Analyst*, **77**, 445–453.
23. Buchala, A. J., Fraser, C. G. and Wilkie, K. C. B. (1972). *Phytochemistry*, **11**, 1249–1254.
24. Carpenter, K. J. and Clegg, K. M. (1956). *J. Sci. Food Agric.*, **7**, 45–51.
25. Castagne, A. E. and Siddiqui, I. R. (1965). *Carbohyd. Res.*, **42**, 382–386.
26. Clegg, K. M. (1956). *J. Sci. Food Agric.*, **7**, 40–44.
27. Cole, E. W. (1967). *Cereal Chem.*, **44**, 411–416.
28. Cole, E. W. (1970). *Cereal Chem.*, **47**, 696–699.
29. Crowell, E. P. and Burnett, B. B. (1967). *Anal. Chem.*, **39**, 121–124.
30. Cundall, R. B., Phillips, G. D. and Rowlands, D. P. (1973). *Analyst*, **98**, 857–862.
31. Cutter, E. G. (1971). *Plant Anatomy—Experiment and Interpretation Part 2, Organs*, Edward Arnold, London.
32. Dako, D. Y., Trautner, K. and Somogyi, J. C. (1970). *Schweiz. Med. Wochschr.*, **100**, 877–903.
33. Dawson, R. M. C., Elliot, D. C., Elliot, W. H. and Jones, K. M. (1969). *Data for Biochemical Research*, 2nd edition, Oxford University Press, London.
34. Dekker, R. F. H. and Richards, G. N. (1972). *J. Sci. Food Agric.*, **23**, 475–483.
35. Dische, Z. (1955). In *Methods of Biochemical Analysis* (Glick, D., ed.), vol. II, pp. 313–358, Interscience Publishers, New York.
36. Dougall, H. W. (1958). *J. Sci. Food Agric.*, **9**, 1–7.
37. Dutton, G. G. S. (1973). *Adv. Carbohyd. Chem.*, **28**, 11–160.
38. Ewald, C. M. and Perlin, A. S. (1959). *Canad. J. Chem.*, **37**, 1254.
39. Fairbridge, R. A., Willis, K. J. and Booth, R. G. (1951). *Biochem. J.*, **49**, 423–427.
40. Ferrier, R. J. and Collins, P. M. (1972). *Monosaccharide Chemistry*, Penguin Books, London.
41. Floridi, A. (1971). *J. Chromatogr.*, **59**, 61–70.
42. Food and Agriculture Organisation of the United Nations (1947). *Energy yielding components of food and computation of calorie values*, FAO, Washington.
43. Fraser, J. R., Brendon-Bravo, M. and Holmes, D. C. (1956). *J. Sci. Food Agric.*, **7**, 577–589.
44. Friedmann, T. E., Witt, N. F., Neighbors, B. W. and Weber, C. W. (1967). *J. Nutr.*, **91**, suppl. 2.
45. Gaillard, B. D. E. (1958). *J. Sci. Food Agric.*, **9**, 170–177.
46. Gaillard, B. D. E. (1961). *Nature*, **191**, 1295–1296.
47. Garcia, W. J. and Wolf, M. J. (1972). *Cereal Chem.*, **49**, 298–306.

48. Glicksman, M. (1962). *Adv. Food Res.*, **11**, 109–200.
49. Goering, H. K. and Van Soest, P. J. (1970). *Forage Fiber Analysis* (*Agric. Handbook no.* 379), U.S. Dept. Agriculture.
50. Greenwood, C. T. (1970). In *The Carbohydrates, Chemistry and Biochemistry* (Pigman, W. and Horton, D., eds), vol. IIB, pp. 471–513, Academic Press, London and New York.
51. Hall, J. and Tucker, D. (1968). *Anal. Biochem.*, **26**, 12–17.
52. Hallab, A. H. and Epps, E. A. (1963). *J. Ass. Off. Agric. Chem.*, **46**, 1006–1010.
53. Hallsworth, E. G. (1950). *Agricultural Progress*, **25**, 39.
54. Henneberg, W. and Stohmann, F. (1860). *Beiträge zur Begründung einer rationellen Fütterung der Wiederkaüer* I, Braunschweig.
55. Hough, L. and Jones, J. K. N. (1962). In *Methods in Carbohydrate Chemistry* (Whistler, R. L. and Wolfrom, M. L., eds), vol. 1, pp. 21–31, Academic Press, New York and London.
56. Howling, D. (1974). *Food Technology in Australia*, **26**, 464–473.
57. IUPAC (1971). *Biochem. J.*, **125**, 673–695.
58. Jacobs, M. B. (1958). *The Chemical Analysis of Foods and Food Products*, 3rd edition, Van Nostrand Co. Inc., Princeton.
59. Jones, J. K. N. and Hay, G. W. (1972). In *The Carbohydrates, Chemistry and Biochemistry* (Pigman, W. and Horton, D., eds), vol. IA, pp. 403–422, Academic Press, New York and London.
60. Jonsson, P. and Samuelson, O. (1967). *Anal. Chem.*, **39**, 1156–1158.
61. Kefford, J. F. (1959). *Adv. Food Res.*, **9**, 285–372.
62. Kent-Jones, D. W. and Amos, A. J. (1967). *Modern Cereal Chemistry*, 6th edition, Food Trade Press Ltd, London.
63. Kesler, R. B. (1967). *Anal Chem.*, **39**, 1416–1422.
64. Knee, M. (1973). *Phytochemistry*, **12**, 1543–1549.
65. Kulka, R. G. (1956). *Biochem. J.*, **63**, 542–548.
66. Lederer, E. and Lederer, M. (1957). *Chromatography—Review of Principles and Applications*, pp. 237–287, Elsevier, Amsterdam and New York.
67. Lee, C. Y., Shallenberger, R. S. and Vittum, M. T. (1970). *Free Sugars in Fruits and Vegetables*, New York State Agriculture Experimental Station, Food Sciences: Food Science and Technology No. 1.
68. Lintas, C. and D'Appolonia, B. L. (1972). *Cereal Chem.*, **49**, 731–736.
69. Long, C. (1961). *Biochemists Handbook*, E. & F. N. Spon Ltd, London.
70. MacDonald, D. L. (1972). In *The Carbohydrates, Chemistry and Biochemistry* (Pigman, W. and Horton, D., eds), vol. IA, pp. 253–277, Academic Press, New York and London.
71. Macrae, J. C. and Armstrong, D. G. (1968). *J. Sci. Food Agric.*, **19**, 578–581.
72. McCance, R. A. and Lawrence, R. D. (1929). *The Carbohydrate Content of Foods, Spec. Rep. Ser. Med. Res. Coun. Lond.* No. 195, HMSO, London.
73. McCance, R. A. and Widdowson, E. M. (1960). *The Composition of Foods, Spec. Rep. Ser. Med. Res. Coun. Lond.* No. 297, HMSO, London.
74. McCance, R. A., Widdowson, E. M. and Shackleton, L. R. B. (1936). *The nutritive value of fruits, vegetables and nuts, Spec. Rep. Ser. Med. Res. Coun. Lond.* No. 213, HMSO, London.
75. McCollum, E. V. (1957). *A History of Nutrition*, Houghton Mifflin Co., Boston.

76. McConnell, A. A. and Eastwood, M. A. (1974). *J. Sci. Food Agric.*, **25**, 1451–1456.
77. McDowell, R. H. (1975). *Chem. & Ind.*, 391–395.
78. Manners, D. J. (1974). In *Essays in Biochemistry* (Campbell, P. N. and Dickens, F., eds), vol. 10, pp. 37–71, Academic Press, London.
79. Maynard, L. A. (1944). *J. Nutr.*, **28**, 443–452.
80. Medcalf, D. G. and Cheung, P. W. (1971). *Cereal Chem.*, **48**, 1–8.
81. Merck Index (1968). 8th edition (Stecher, P. G., ed), Merck & Co. Inc., Rahway.
82. Mühlethaler, K. (1961). In *The Cell* (Brachet, J. and Minsky, A. E., eds), vol. 2, pp. 85–134, Academic Press, New York.
83. Mopper, K. (1972). *Anal. Biochem.*, **45**, 147–153.
84. Morley, R. G., Phillips, G. O., Power, D. M. and Morgan, R. E. (1972). *Analyst*, **97**, 315–319.
85. Neukom, H., Providoli, L., Gremli, H. and Hui, P. A. (1967). *Cereal Chem.*, **44**, 238–244.
86. Norman, A. G. (1937). *The Biochemistry of Cellulose, the Polyuronides, etc.*, Oxford University Press, London.
87. O'Dwyer, M. H. (1926). *Biochem. J.*, **20**, 656–664.
88. Percival, E. (1963). In *Comprehensive Biochemistry* (Florkin, M. and Stotz, E. H., eds), vol. 5, pp. 1–66, Elsevier, Amsterdam, London and New York.
89. Percival, E. (1970). In *The Carbohydrates, Chemistry and Biochemistry* (Pigman, W. and Horton, D., eds), vol. IIB, pp. 537–568, Academic Press, New York and London.
90. Pierce Chemical Company (1970). *Handbook of Silylation*, Handbook GPA-3, Pierce Chemical Co., Rockford.
91. Pigman, W. W. and Goepp, R. M. Jr (1948). *Chemistry of the Carbohydrates*, pp. 139–142, Academic Press, New York.
92. Pigman, W. W. and Horton, D. (1972). *The Carbohydrates, Chemistry and Biochemistry*, 2nd edition, vol. IA, Academic Press, New York.
93. Pritchard, P. J., Dryburgh, E. A. and Wilson, B. J. (1973). *J. Sci. Food Agric.*, **24**, 663–668.
94. Pucher, C. W., Leavenworth, C. S. and Vickery, H. B. (1948). *Anal. Chem.*, **20**, 850–853.
95. Robyt, J. F., Ackerman, R. J. and King, J. G. (1972). *Anal. Biochem.*, **45**, 517–524.
96. Roe, J. H. (1955). *J. Biol. Chem.*, **212**, 335–343.
97. Selvendran, R. R. (1975). *Phytochemistry*, **14**, 1011–1017.
98. Shafizadeh, F. and McGinnis, G. D. (1971). *Adv. Carbohyd. Chem.*, **26**, 297–349.
99. Siegel, S. M. (1968). In *Comprehensive Biochemistry* (Florkin, M. and Stotz, E. H., eds), vol. 26A, pp. 1–51, Elsevier, Amsterdam.
100. Smith, F. and Montgomery, R. (1959). *The Chemistry of Plant Gums and Mucilages*, Rheinhold, New York.
101. Somogyi, M. (1945). *J. Biol. Chem.*, **160**, 61–68.
102. Somogyi, J. C. and Trautner, K. (1974). *Schweiz. Med. Wochschr.*, **104**, 177–182.
103. Southgate, D. A. T. (1969a). *J. Sci. Food Agric.*, **20**, 326–329.

104. Southgate, D. A. T. (1969b). *J. Sci. Food Agric.*, **20**, 331–335.
105. Southgate, D. A. T. (1973). In *Nutritional Problems in a Developing World* (Hollingsworth, D. and Russell, M., eds), pp. 199–204, Applied Science Publishers, London.
106. Southgate, D. A. T. (1974). *J. Assoc. Pub. Analysts*, **12**, 114–118.
107. Southgate, D. A. T. (1975). *Biblthca Nutr. Dieta*, **22**, 109–124.
108 Southgate, D. A. T. and Durnin, J. V. G. A. (1970). *Br. J. Nutr.*, **24**, 517–535.
109. Spector, W. S. (1956). *Handbook of Biological Data*, W. B. Saunders & Co. Ltd, Philadelphia.
110. Stahl, E. and Kallenbach, U. (1965). *Thin-layer Chromatography: a Laboratory Handbook* (Stahl, E., ed.), pp. 461–469, Springer-Verlag, Berlin, Heidelberg and New York.
111. Statutory Instruments (1972). *Labelling of Food Regulations* 1970, SI 400, p. 1510, HMSO, London.
112. Sweeley, C. C., Bentley, R., Makita, M. and Wells, W. W. (1963). *J. Amer. Chem. Soc.*, **85**, 2497–2507.
113. Terry, R. A. and Outen, G. E. (1973). *Chem. & Ind.*, **23**, 1116–1117.
114. Timell, T. E. (1964). *Adv. Carbohyd. Chem.*, **19**, 247–307.
115. Timell, T. E. (1965). *Adv. Carbohyd. Chem.*, **20**, 410–483.
116. Tracey, M. V. (1948). *Biochem. J.*, **43**, 185–189.
117. Trautner, K. and Somogyi, J. C. (1972). *Trav. Chim. Ali. Hyg.*, **63**, 240–260.
118. Trowell, H. (1972). *Amer. J. Clin. Nutr.*, **25**, 926–932.
119. Van Soest, P. J. (1963a). *J. Ass. Off. Agric. Chem.*, **46**, 825–829.
120. Van Soest, P. J. (1963b). *J. Ass. Off. Agric. Chem.*, **46**, 829–835.
121. Van Soest, P. J. and McQueen, R. W. (1973). *Proc. Nutr. Soc.*, **32**, 123–130.
122. Van Soest, P. J. and Wine, R. H. (1968). *J. Ass. Off. Agric. Chem.*, **51**, 780–785.
123. Waite, R. and Gorrod, A. R. N. (1959a). *J. Sci. Food Agric.*, **10**, 308–317.
124. Waite, R. and Gorrod, A. R. N. (1959b). *J. Sci. Food Agric.*, **10**, 317–326.
125. Walborg, E. F. Jr, Ray, D. B. and Öhrebearg, L. E. (1969). *Anal. Biochem.*, **29**, 433–440.
126. Walker, D. M. and Hepburn, W. R. (1955). *Agricultural Progress*, **30**, 118–119.
127. Wardrop, A. B. (1962). *Botan. Rev.*, **28**, 241–285.
128. Whalley, H. C. S. (Ed.) (1964). *Methods of Sugar Analysis*, Int. Comm. Uniform Meth. Sugar Analysis, Elsevier, Amsterdam, London and New York.
129. Whistler, R. L. and Feather, M. S. (1965). In *Methods in Carbohydrate Chemistry* (Whistler, R. L., ed.), vol. V, pp. 144–145, Academic Press, New York and London.
130. Whistler, R. L. and Wolfrom, M. L. (1962). *Methods in Carbohydrate Chemistry*, vol. I, Academic Press, New York and London.
131. Whistler, R. L., Bachrach, J. and Bowman, D. R. (1948). *Arch. Biochem.*, **18**, 25–33.
132. Widdowson, E. M. (1955). *Proc. Nutr. Soc.*, **14**, 142–148.
133. Widdowson, E. M. and McCance, R. A. (1935). *Biochem. J.*, **29**, 151–156.
134. Williams, R. D. and Olmsted, W. H. (1935). *J. Biol. Chem.*, **108**, 653–666.
135. Williams, R. D., Wicks, L., Bierman, H. R. and Olmsted, W. H. (1940). *J. Nutr.*, **19**, 593–604.
136. Wylie, A. (1973). *J. Roy. Soc. Health*, **6**, 309–315.

Index